SPATIAL VARIATION AND SEASONALITY IN GROWTH AND REPRODUCTION OF *ENHALUS ACOROIDES* (L.f.) ROYLE POPULATIONS IN THE COASTAL WATERS OFF CAPE BOLINAO, NW PHILIPPINES

Spatial variation and seasonality in growth and reproduction of *Enhalus acoroides* (L.f.) Royle populations in the coastal waters off Cape Bolinao, NW Philippines

DISSERTATION
Submitted in fulfilment of the requirements of
the Board of Deans of Wageningen Agricultural University
and the Academic Board of the International Institute for Infrastructural,
Hydraulic and Environmental Engineering for the Degree of DOCTOR
to be defended in public
on Tuesday, 10 March 1998 at 16:00 h in Delft

by

RENÉ NADÁL ROLLÓN
born in Bohol, Philippines

A.A. BALKEMA / ROTTERDAM / BROOKFIELD / 1998

This dissertation has been approved by the promoters:
Dr W. van Vierssen, professor in Aquatic Ecology
Dr W.J. Wolff, professor in Marine Biology

Co-promoter:
Dr E. D. de Ruyter van Steveninck, senior scientist at the department of Environmental Science and Water Resources Management, IHE

Published by
A.A. Balkema, P.O. Box 1675, 3000 BR Rotterdam, Netherlands
Fax: +31.10.4135947; E-mail: balkema@balkema.nl; Internet site: http://www.balkema.nl

A.A. Balkema Publishers, Old Post Road, Brookfield, VT 05036-9704, USA
Fax: 802.276.3837; E-mail: info@ashgate.com

ISBN 90 5410 412 0

Printed in the Netherlands

Malthusian overfishing occurs when poor fishermen, faced with declining catches and lacking any other alternative, initiate wholesale resource destruction in their effort to maintain their incomes. This may involve in order of seriousness, and generally in temporal sequence: (1) use of gears and mesh sizes not sanctioned by the government; (2) use of gears not sanctioned within the fisherfolk communities and/or catching gears that destroy the resource base; and (3) use of "gears" such as dynamite or sodium cyanide that do all of the above and even endanger the fisherfolks themselves.

(from Pauly et al. 1989
as qouted in McManus et al. 1992)

Alang sa akong inaháng pulô
(nga nagsubô tungód sa Malthusian overfishing):

Sa daghang mgá katuigan nga akó walâ sa imong baybayon,
ubán kanunay kanimo ang akong pangisip ug kasingkasing.
Kun, dinhà sa imong hunasan, mabanhaw lamang untà,
ang napukan nga nagkadaiyang matang nga kinabuhì,
nga matá'g usá, sakop sa dilì matukib nga laing kalibutan,
walâ na'y dapít, mabisan-asa pa man,
nga molabáw sa imong kaanyag ug kaadunahan!

Contents

Preface

This PhD research was financed by the Environmental Ecotechnology programme, an international cooperation between The Netherlands, The Philippines and Uganda, aiming at executing an integrated environmental research in aquatic ecosystems. Prof. Dr. Wim van Vierssen initiated this international cooperation and made it a reality. At a later stage, Dr. Erik de Ruyter van Steveninck assumed the task of coordinating towards the successful execution of various components. In particular, this PhD research benefited from the facilities of a number of institutions, namely the Marine Science Institute (CS, UP Diliman, Philippines), the IHE (Delft, The Netherlands) and the NIOO-CL (Nieuwersluis, The Netherlands). This PhD research also benefited from the cooperation with Paul Rivera who also conducted a PhD research at the same period and study area within the context of Environmental Ecotechnology programme.

Operationalizing the general framework of this research had been, in various points, difficult. Many thanks to Prof. Wim van Vierssen, Dr. Erik de Ruyter van Steveninck and Dr. Jan Vermaat whose combined critical ideas were, in all cases, enlightening.

I would like to express my gratitude to the Marine Science Institute of the University of the Philippines, in general, and to Dr. Miguel D. Fortes, in particular, firstly, for giving me the 'breaks' while being a member of his research team (1988-1992), and, secondly, for kindly making available every facility I needed during the data collection (1993-1996). To all friends and colleagues at the Institute who together provided me a scientific home, ... many thanks. The list of your names is just too long for a preface page. However, the "seagrass people" cannot be left unmentioned. To Helen D. and Jean B., thank you for the administrative assistance.

The collection of field data would not have been possible without the magnificient assistance of Jack Rengel. Aside from being a supportive diving buddy, he always masterfully piloted the boat to and from the study sites, through the tricky maze inside the reef flat. My acknowledgement also goes to Dillard Knight for doing his MSc research within the context of this dissertation.

The writing of this dissertation was mostly done at the Netherlands Institute of Ecology - Center for Limnology (NIOO-CL). My gratitude goes to all the staffmembers who, in one way or another, made my writing possible. The place itself is wonderful. Maraming salamat!

To Analiza and Elise, you are a marvelous team! In you, I always found the ultimate spirit to go on at various frustrating moments.

Finally, I'd like to thank my parents Enrique and Esperanza. Inyo usáb kiní nga kalampusan. Ang unang lakang sa akong halayóng panaw nagsugod dinhà kaninyo.

René N. Rollón
September 1997

Chapter 1

General introduction

Seagrasses are the only submerged flowering plants in the marine environment. They all belong to one order (*Helobiae*) under the monocotyledon group (i.e., no marine dicotyledon exists) and the total number of species so far recorded is less than 60 (57, Kuo & McComb 1989; 49, Den Hartog 1970), a negligible number (Den Hartog 1970) compared to the number of flowering species found in the terrestrial environment. However, when present, seagrasses and the extensive meadows they can form play an important role in the marine environment. This can be summarized in six basic axioms (Larkum et al. 1989): (1) stability of structure, (2) provision of food and shelter for many organisms, (3) high productivity, (4) recycling of nutrients, (5) stabilizing effects on shorelines and (6) provision of a nursery ground for various fauna including those which are of commercial importance.

As primary requirements to photosynthetic organisms, light and mineral nutrition affect seagrasses. Thus, environmental changes, e.g. erosion-derived siltation (EMB Report 1990; Yap 1992; Vermaat et al. in press), which leads to light climate deterioration could spell serious threat to seagrass meadows. In fact, human activities leading to the reduction of water clarity (which includes nutrient loading, eutrophication, water quality, pollution, and turbidity) ranked top on the worldwide list of causes of seagrass declines and disappearances (Short & Wyllie-Echeverria 1996). Because seagrasses need nutrition, an increase in nutrients in the environment, per se, may enhance seagrass productivity. However, excessive nutrient increases also lead to periphyton and phytoplankton dominance which in effect deprives seagrasses of light. As a net effect, loss of seagrasses is likely to happen (e.g. Cambridge et al. 1986). Aside from environmental changes due to anthropogenic circumstances, seagrasses also have to respond to natural phenomena, e.g. 'wasting' disease, (Rasmussen 1977; Den Hartog 1987; Roblee et al. 1991), cyclones, (Cambridge 1975; Poiner et al. 1989) and physical clearances by, for instance, dugongs (De Iongh 1995; Preen 1995). Particularly for the highly diverse SE Asian seagrass meadows, our understanding of seagrass responses to environmental changes, whether natural- or man-induced, is poor.

In the tropical Indo-West Pacific Region, *Enhalus acoroides* (L.f.) Royle is the largest species (Den Hartog 1970; Meñez et al. 1983; Kuo & McComb 1989; Fortes 1988) among the total of ca. 20 species occurring, and it structurally dominates many seagrass meadows in the region (Erftemeijer 1993; Tomasko et al. 1993; Fortes 1994; Japar 1994; Kiswara 1994; Loo et al. 1994; Poovachiranon et al. 1994; Vermaat et al. 1995), often together with *Thalassia hemprichii* (Ehrenb.) Aschers. (Erftemeijer et al. 1993; Vermaat et al. 1995; pes. obs.; see also the appendix summarizing the taxonomy and distribution of these two most common species). Thus, it might be argued that understanding the biology of, especially, *Enhalus acoroides*, could contribute substantially to understanding the effects of environmental changes on seagrass meadows.

It is for the above argument that this dissertation focuses on *Enhalus acoroides*. The main objective of this dissertation is to gain insight into the general question of "how this species would respond to environmental changes such as increased turbidity, nutrient loading and siltation". To approach this broad research question, there is a need to gather information on (1) the growth characteristics of *Enhalus* in the established phase in relation to the prevailing environmental conditions, (2) the influence of these characteristics on its capacity to vegetatively and sexually reproduce, and (3) the fitness of the propagules (produced by reproduction) in response to such environmental changes. These three basic research subtopics form the framework of this dissertation.

In the period June 1993-April 1996, a research program based on these lines was carried out in Bolinao, NW Philippines. The coral reef system in the area is largely comprised of a reef flat where seagrasses abound (McManus et al. 1992), often in mixed-species meadows (Vermaat et al. 1995) but also in monospecific stands of *Enhalus acoroides* (pers. obs.). In Bolinao, the Marine Science Institute of the University of the Philippines has a marine laboratory where basic facilities are available to conduct semi-controlled outdoor experiments as well as to access these seagrass meadows and perform *in situ* studies. Various sites with and without *Enhalus* were selected. Of the sites with *Enhalus*, further site selection was made on the basis of, initially, visual differences in (1) shoot size and abundance, (2) presence of other seagrass species, and (3) environmental conditions, e.g. sediment type and turbidity. In most aspects in the biology of *Enhalus acoroides* addressed here, differences in light availability appeared to be the most important. Hence, a number of experiments manipulating light were done. The effects of sediment type and nutrients were also tested experimentally.

The next six chapters report on the results of various studies during the research period. Although this dissertation concentrates its efforts on *Enhalus acoroides*, some aspects of the species biology are best understood when compared with other co-existing species. Hence, the similarly abundant *Thalassia hemprichii* was included for the growth studies. In the aspect of vegetative reproduction, the inclusion of other co-existing species, *Syringodium isoetifolium* (Aschers.) Dandy, *Halophila ovalis* (R. Br.) Hook f., *Halodule uninervis* (Forssk.) Aschers., *Cymodocea rotundata* Ehrenb. & Hempr. ex Aschers., and *Cymodocea serrulata* (R. Br.) Aschers. & Magnus was likewise done.

Immediately succeeding this chapter is the characterization of the relevant environmental conditions at the selected sites (Chapter 2). Chapter 3 deals with shoot size and growth of *Enhalus acoroides* and the co-abundant *Thalassia hemprichii*, focusing on major factors controlling the spatio-temporal variation. Chapter 4 characterizes the recolonization capacities, not only of *Enhalus*, but also, of all other co-existing species in response to artificial disturbance. The chapter on the sexual reproduction of *Enhalus acoroides* (Chapter 5) completes the descriptive aspects of this dissertation. Results of *in situ* and laboratory experiments testing the effects of some relevant factors (based on the results obtained from chapters 2, 3, 4 and 5) are presented in chapters 6 and 7. Chapter 6 reports on the effects of sediment type, shading and nutrient addition on the performance of *Enhalus* seedlings cultured in the laboratory vis-a-vis the establishment, survival and shoot sizes of seedlings grown *in situ*. For mature stands of *Enhalus acoroides*, chapter 7 describes the results of *in situ* shading and transplantation experiments. A general discussion concludes this present research and also speculates on the implications of changes in primary environmental factors for different aspects of the *Enhalus acoroides* biology.

Appendix: Taxonomy and distribution of *Enhalus acoroides* (L.f.) Royle and *Thalassia hemprichii* (Ehrenb.) Aschers.

A number of publications are available describing rather comprehensively the taxonomy, distribution and the general ecology of *Enhalus acoroides* and *Thalassia hemprichii*. These publications include the works of Den Hartog (1970), Meñez et al. (1983), Fortes (1986) and Phillips & Meñez (1988). The following brief summary was extracted from these previous works. In some cases, quantitative descriptions were modified to agree with personal observations and the results obtained from this thesis.

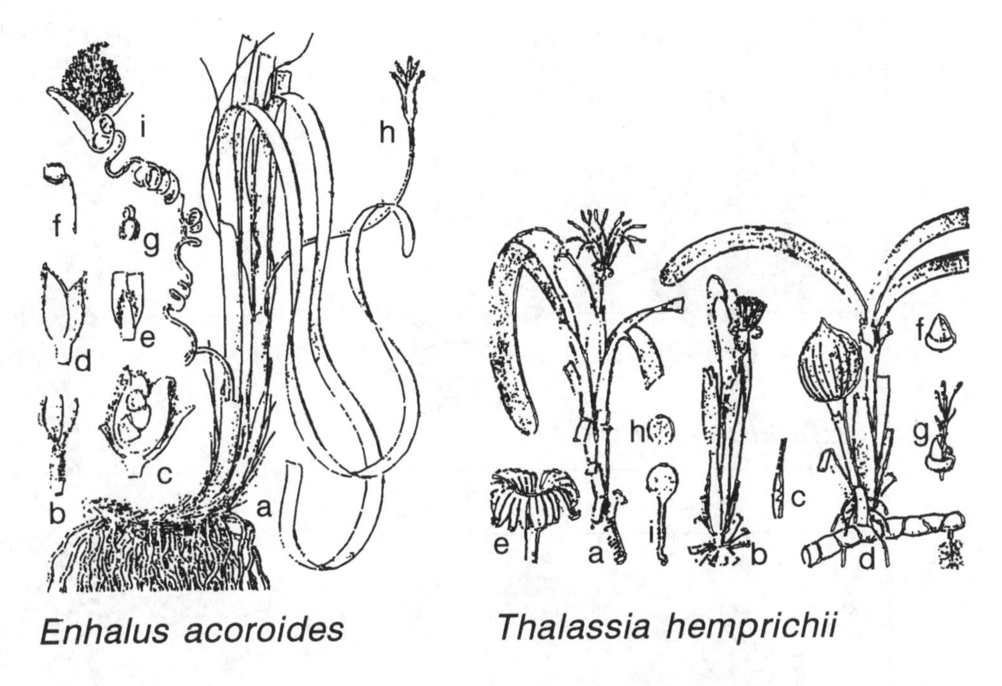

Fig. 1.1. Habitus of *Enhalus acoroides* (L.f.) Royle and *Thalassia hemprichii* (Ehrenb.) Aschers. after Den Hartog (1970): *Enhalus acoroides* - **a**. habitus, **b**. female flower, **c**. longitudinal section of fruit, **d**. male spathe, **e**. longitudinal section of a male spathe, **f**. male flower bud, **g**. opened male flower, **h**. female flower (unfertilized), **i**. fruit on a curled peduncle; *Thalassia hemprichii* - **a**. flowering female plant, **b**. flowering male plant, **c**. ovary after fertilization, **d**. fruiting plant, **e**. opened fruit, **f**. seed, **g**. seedling, **h**. mature pollen grain, **i**. germinating pollen grain.

Both *Enhalus acoroides* and *Thalassia hemprichii* (Fig. 1.1) belong to the same family Hydrocharitaceae which, in contrast to the other family (Potamogetonaceae) found in the marine environment, is best distinguished by the absence of a ligula. Both species are dioecious, bearing pedunculate inflorescences, and, relative to other tropical Indo-West Pacific species, produce probably the largest fruits (ovoid, 5-7 cm long, *Enhalus acoroides*; globose, 2-2.5 cm long, 1.75-3.25 cm wide, *Thalassia hemprichii*) . In both species, the seeds (obconical, 1-1.5 cm, *Enhalus acoroides*; conical, 8 mm, *Thalassia hemprichii*) are released when the ripe fruit bursts.

In comparison, *Enhalus acoroides* is much larger than *Thalassia hemprichii*. Its leaves are wider (12.5-20 mm vs. 4-11 mm) and taller (30-150 cm vs. 10-40 cm). The rhizome of *Enhalus acoroides* is also massive (1.5 cm in diameter vs. 0.3-0.5 cm in *Thalassia hemprichii),* branching monopodially when a new shoot is formed. In contrast, new shoots of *Thalassia hemprichii* stem vertically from the creeping rhizome as short lateral branches. *Enhalus acoroides* has no lateral branches. Roots of *Enhalus acoroides* are coarse, cordlike and unramified, ca.

10-30 cm long and 3-5 mm thick arising from the axillary buds of the ventral leaves. Roots of *Thalassia hemprichii* are also unbranched but densely set with fine hairs, each root arising from nodes of the creeping rhizome (internodes 4-7 mm long). The male *Enhalus acoroides* bears a single pedunculate inflorescence containing numerous flowers (ca. 531 ± 39, n = 3; pers. obs.), each is highly reduced in form to a small free floating device. The male *Thalassia hemprichii* bears 1-2 pedunculate and uniflorous inflorescences. Female plants of both species bear single uniflorous inflorescences, but the peduncle of a female *Enhalus* flower is much longer (30-150 cm vs. 1-1.5 cm in *T. hemprichii*). Pollination in *Enhalus acoroides* occurs at the water surface, whereas that in *Thalassia hemprichii* does so underwater.

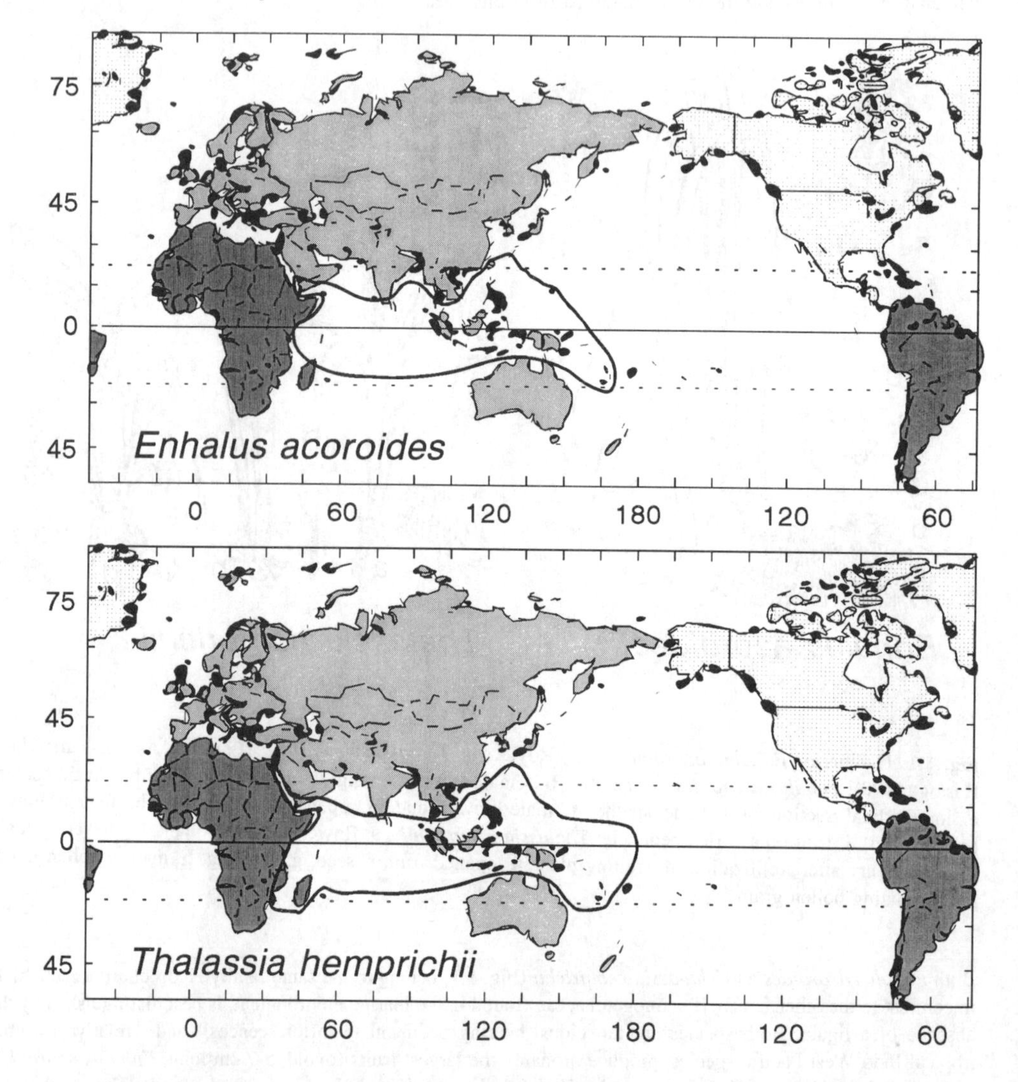

Fig. 1.2. The geographical distribution of *Enhalus acoroides* (L.f.) Royle and *Thalassia hemprichii* (Ehrenb.) Aschers. after Den Hartog (1970) and Phillips & Meñez (1988).

In contrast to these differences in morphology and floral biology, the geographical distribution range of *Enhalus acoroides* and *Thalassia hemprichii* is closely similar (Fig. 1.2) with the latter extending its longitudinal range only a bit wider (East, as far as 180° meridian; West, Red Sea to Mozambique; Mwaiseje (1979) reported an

Enhalus acoroides

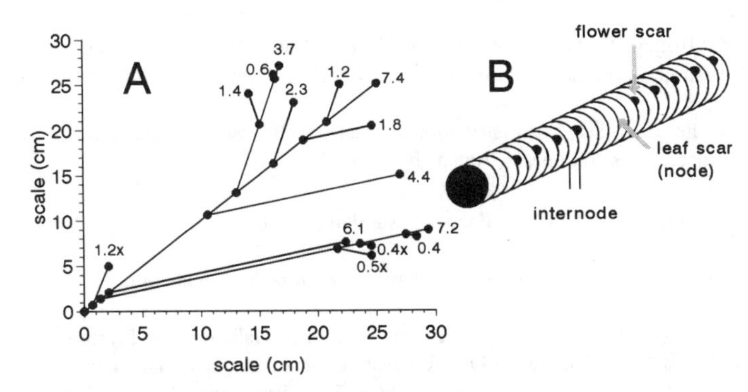

Thalassia hemprichii

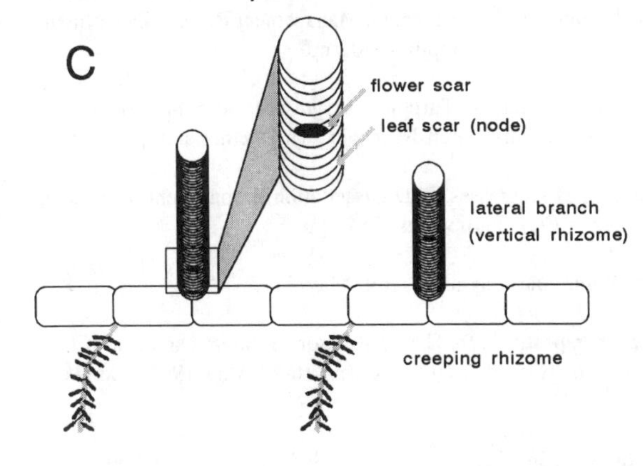

Fig. 1.3. A: Scaled diagram of an *Enhalus acoroides* clone (shoot cluster no. 3, site 6, see Chapters 2 & 5). Lines represent lengths of the monopodially branching rhizomes (straight line = shoot; branch = new shoot). Numbers at the terminal ends represent shoot ages (in years) at the time of collection; x = dead shoot. **B**: Diagram of an *Enhalus acoroides* rhizome segment showing leaf and flower scars and internodes (ca. 30-day growth equivalents, Rollon et al., Unpubl. data). **C**: Schematic diagram of *Thalassia hemprichii* growth form similarly showing leaf and flower scars and vertical rhizome internodes (ca. 10-day growth equivalents: Vermaat et al. 1995; Rollon et al., Unpbl. data).

Enhalus acoroides occurrence on the coast of Tanzania). Both are widely distributed in the tropical parts of the Indian Ocean and the Western Pacific and are very common in the Indo-Malay Archipelago and in the Philippines.

In both species, shed leaves and flowers leave scars on their respective rhizomes (vertical rhizomes for *Thalassia hemprichii*; monomorphic rhizome for *Enhalus acoroides*; Fig. 1.3; cf. Tomlinson 1974). The horizontal (creeping) rhizomes in *Thalassia hemprichii* bear reduced scaly leaves. This architecture allows the quantification of rhizome growth and flowering events once the corresponding plastochrone intervals have been determined. The details of this age reconstruction technique which proved powerful in describing seagrass growth dynamics and flowering events over long-term scale have been formalized by Duarte et al. (1994).

LITERATURE CITED

Cambridge, M.L. 1975. Seagrasses of south-western Australia with special reference to the ecology of *Posidonia australis* Hook. *f.* in a polluted environment. Aquat. Bot. 1: 149-161.

Cambridge, M.L., A.W. Chiffings, C. Brittan, L. Moore and A.J. McComb. 1986. The loss of seagrass in Cockburn Sound, Western Australia 2. Possible causes of seagrass decline. Aquat. Bot. 24: 269-285.

De Iongh, H.H., B.J. Wenno and E. Meelis. 1995. Seagrass distribution and seasonal biomass changes in relation to dugong grazing in the Moluccas, East Indonesia. Aquat. Bot. 50: 1-19.

Den Hartog, C. 1970. The seagrasses of the world. North Holland, Amsterdam, 275 pp.

Den Hartog, C. 1987. 'Wasting disease' and other dynamic phenomena in *Zostera* beds. Aquat. Bot. 27: 3-14.

Duarte, C.M., N. Marba, N. Agawin, J. Cebrian, S. Enriquez, M.D. Fortes, M.E. Gallegos, M. Merino, B. Olesen, K. Sand-Jensen, J. Uri and J. Vermaat. 1994. Reconstruction of seagrass dynamics: age determinations and associated tools for the seagrass ecologist. Mar. Ecol. Prog. Ser. 107: 195-209.

EMB Report. 1990. The Philippine environment in the eighties. Environmental Management Bureau, Department of Environment and Natural Resources, Quezon City, Philippines, 302 pp.

Erftemeijer, P.L.A. 1993. Factors limiting growth and production of tropical seagrasses: nutrient dynamics in Indonesian seagrass beds. PhD thesis, Katholieke Universiteit Nijmegen, The Netherlands, 173 pp.

Fortes, M.D. 1986. Taxonomy and ecology of Philippines seagrasses. PhD Dissertation, Department of Botany, University of the Philippines, Diliman, Quezon City, Philippines, 254+ pp.

Fortes, M.D. 1988. Indo-West Pacific affinities of Philippine seagrasses. Bot. Mar. 31: 237-242.

Fortes, M. D. 1994. Philippine seagrasses: status and perspectives. In: C.R. Wilkinson, S. Sudara and L.M. Chou (eds.). Proc. 3rd. ASEAN-Australia Symp. on living coastal resources, 16-20 May 1994, Bangkok, Thailand, 291-310 pp.

Japar, S.B. 1994. Status of seagrass resources in Malaysia. In: C.R. Wilkinson, S. Sudara and L.M. Chou (eds.). Proc. 3rd. ASEAN-Australia Symp. on living coastal resources, 16-20 May 1994, Bangkok, Thailand, 283-289 pp.

Kiswara, W. 1994. A review: seagrass ecosystem studies in Indonesian waters. In: C.R. Wilkinson, S. Sudara and L.M. Chou (eds.). Proc. 3rd. ASEAN-Australia Symp. on living coastal resources, 16-20 May 1994, Bangkok, Thailand, 259-281 pp.

Kuo, J. and A.J. McComb. 1989. Seagrass taxonomy, structure and development. In: A.W.D. Larkum, A.J. McComb and S.A. Shepherd (eds.). Biology of seagrasses, a treatise on the biology of seagrasses with special reference to the Australian region. Elsevier, Amsterdam, 6-73 pp.

Larkum, A.W.D., A.J. McComb and S.A. Shepherd (eds.). 1989. Biology of seagrasses, a treatise on the biology of seagrasses with special reference to the Australian region. Elsevier, Amsterdam, 841 pp.

Loo, M.G.K., K.P.P. Tun, J.K.Y. Low and L.M. Chou. 1994. A review of seagrass communities in Singapore. In: C.R. Wilkinson, S. Sudara and L.M. Chou (eds.). Proc. 3rd. ASEAN-Australia Symp. on living coastal resources, 16-20 May 1994, Bangkok, Thailand, 311-316 pp.

Mwaiseje, B. 1979. The occurrence of *Enhalus* on the coast of Tanzania. Aquat. Bot. 7: 393.

McManus, J.W., C.L. Nanola, Jr., R.B. Reyes, Jr. and K.N. Keshner. 1992. Resource ecology of the Bolinao coral reef system. ICLARM Stud. Rev. 22, 117 pp.

Meñez, E.G., R.C. Phillips and H. Calumpong. 1983. Seagrasses from the Philippines. Smithsonian Contrib. Mar. Sci. 21, 40 pp.

Phillips, R.C. and E.G. Meñez. 1988. Seagrasses. Smithsonian Contrib. Mar. Sci. 34, 104 pp.

Poiner, I.R. 1989. Regional studies - Seagrasses of tropical Australia. In: A.W.D. Larkum, A.J. McComb and S.A. Shepherd (eds.). Biology of seagrasses, a treatise on the biology of seagrasses with special reference to the Australian region. Elsevier, Amsterdam, 279- 303 pp.

Poovachiranon, S., S. Nateekanjanalarp and S. Sudara. 1994. Seagrass beds in Thailand. In: C.R. Wilkinson, S. Sudara and L.M. Chou (eds.). Proc. 3rd. ASEAN-Australia Symp. on living coastal resources, 16-20 May 1994, Bangkok, Thailand, 317-321 pp.

Preen, A. 1995. Impacts of dugong foraging on seagrass habitats: observational and experimental evidence for cultivation grazing. Mar. Ecol. Prog. Ser. 124: 201-213.

Rasmussen, E. 1977. The wasting disease of eelgrass (*Zostera marina*) and its effects on environmental factors and fauna. In: C.P. McRoy and C. Helferrich (eds.). Seagrass ecosystems: a scientific perspective, Marcel Decker, New York, 1-51 pp.

Robblee, M.B., T.R. Barber, P.R. Carlson, Jr., M.J. Durako, J.W. Fourqurean, L.K. Muehlstein, D. Porter, L.A. Yarbro, R.T. Zieman and J.C. Zieman. 1991. Mass mortality of the tropical seagrass *Thalassia testudinum* in Florida Bay (USA). Mar Ecol. Prog. Ser. 71: 297-299.

Short, F.T. and S. Wyllie-Echeverria. 1996. Natural and human-induced disturbance of seagrasses. Environmental Conservation 23: 17-27.

Tomasko, D.A., C.J. Dawes, M.D. Fortes, D.B. Largo and M.N.R. Alava. 1993. Observations on a multi-species seagrass meadow off-shore of Negros Occidental, Republic of the Philippines. Bot. Mar. 36: 303-311.

Tomlinson, P.B. 1974. Vegetative morphology and meristem dependence - the foundation of productivity in seagrasses. Aquaculture 4: 107-130.

Vermaat, J.E., N.S.R Agawin, C.M. Duarte, S. Enriquez, M.D. Fortes, N. Marba, J.S. Uri and W. van Vierssen. In press. The capacity of seagrasses to survive eutrophication and siltation, across-regional comparisons. Ambio.

Vermaat, J.E., N.S.R. Agawin, C.M. Duarte, M.D. Fortes, N. Marba and J.S. Uri. 1995. Meadow maintenance, growth and productivity of a mixed Philippine seagrass bed. Mar. Ecol. Prog. Ser. 124: 215-225.

Yap, H. T. 1992. Marine environmental problems - experiences of developing regions. Mar. Pol. Bull. 25: 1-4.

Chapter 2

Characterization of the environmental conditions at the selected study sites: seagrass habitats of Bolinao, NW Philippines

Abstract. In connection with the study of spatio-temporal variation in growth and reproduction of *Enhalus acoroides*, a characterization of the environmental conditions at selected sites was done. A number of environmental parameters differed significantly between sites although generally narrow in magnitude. Sites close to the mainland (inner sites) had higher values than "outer" sites in terms of vertical light attenuation coefficient, sediment organic matter, sediment N & K and periphyton biomass. Of the variables measured, the incident light (PAR), rainfall and temperature showed distinct temporal patterns. Salinity values were fairly constant (28-34 $^{0}/_{00}$) over a year except at a southern site where extreme drops (down to 20 $^{0}/_{00}$) could occur over short periods during rainy seasons. Periphyton colonization and light attenuance (amounting up to 75%) differed across time and sites. An estimation of the absolute PAR reaching seagrass depth at the sites suggested that at the deepest site (5 m), there is probably a serious light limitation for *Enhalus*.

INTRODUCTION

In a preliminary visual survey, a number of sites in Bolinao (NW Philippines) differed in several aspects. Shoot size and density distribution of seagrasses, particularly *Enhalus acoroides* (L.f.) Royle, visually differed across mean water depth and, sites deeper than 3 meters had no *Enhalus*. Some of these *Enhalus* populations occurred in sites near the coast where turbidity was visually higher. This chapter quantifies these visual observations and aims to characterize different sites with and without *Enhalus acoroides*. Temporal and spatial differences in prevailing environmental conditions may then be causally linked to the spatio-temporal variation in different aspects of the biology of *Enhalus acoroides* and other less-dominant seagrasses in the study area (succeeding Chapters 3-7).

Results are categorized into two: physico-chemical and biological factors. The physico-chemical factors measured in this study include vertical light attenuation coefficient, temperature, salinity, sedimentation flux, water flow velocity, sediment grain-size distribution and organic matter content, nutrients N, P, K, rainfall, tides and global radiation. For biological factors, this chapter reports on an overview of seagrass community structures at different sites and quantifies biomass and light attenuance due to periphyton colonizing on artificial seagrass leaves.

STUDY AREA

Bolinao, Pangasinan (Fig. 2.1) is located in the northwestern Philippines, a coastal municipality of about 50,000 people. In a broader view, Bolinao is an integral part of Lingayen Gulf (see also McManus & Chua 1990, McManus et al. 1992, Rivera 1997) where siltation is threatening the entire system (see also Rivera 1997; this dissertation, Chapter 8). The geology of the mainland part of Bolinao is basically limestone with its topsoil producing only limited crops. Agricultural products have only little significance to the town's economy, and being so, the livelihood has been heavily dependent on the harvest from the marine

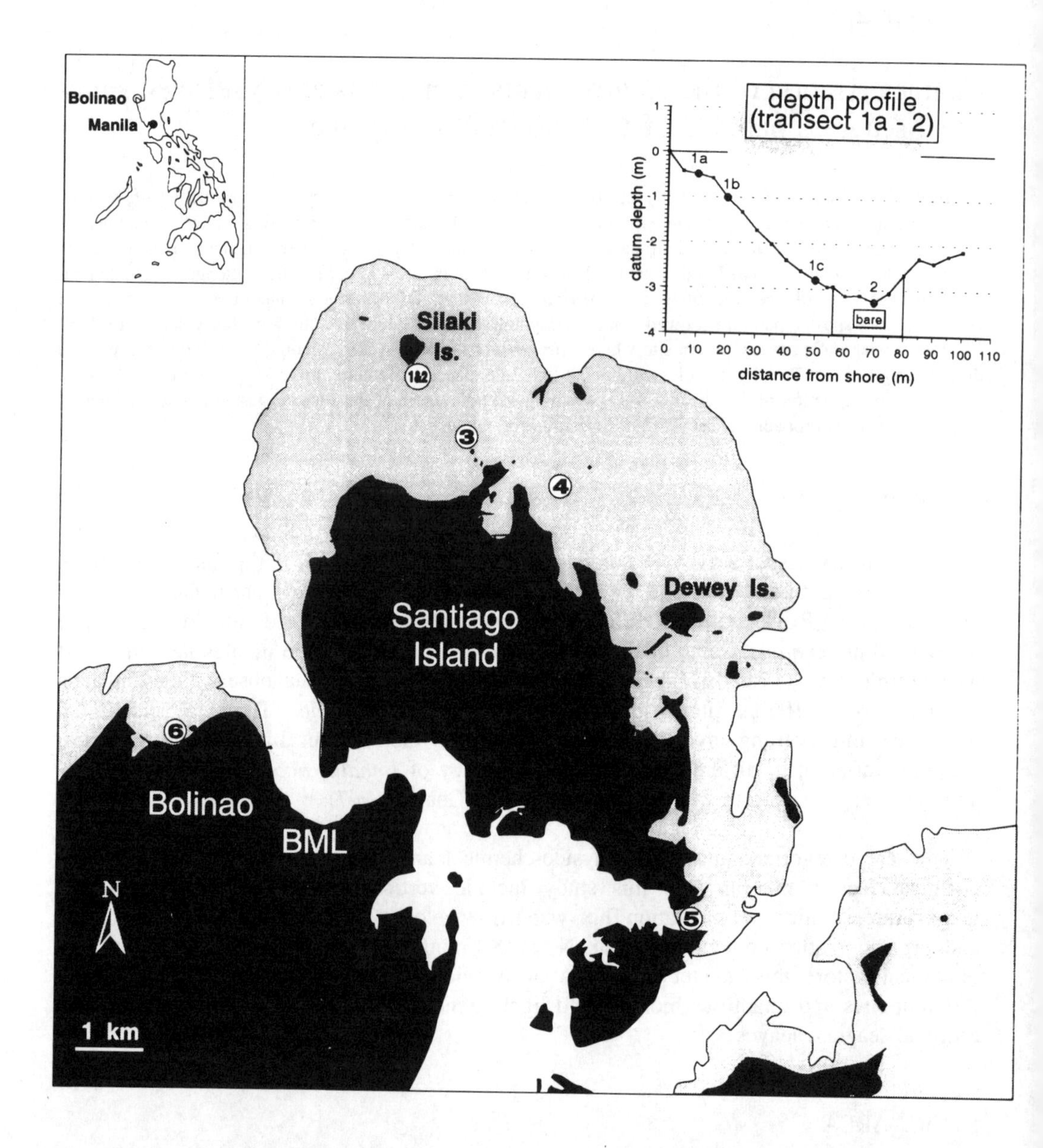

Fig. 2.1. Map showing the study area (Bolinao, Pangasinan, northwestern Philippines). Sites are indicated by numbers 1-6. Sites 1a, 1b and 1c are adjacent sites along a depth gradient (0.5, 1.0 and 3.0 m below zero datum; see inset). Site 2 has no seagrasses (inset, shaded x-axis range). Visually turbid sites 5 and 6 are comparable to 1a in terms of mean water depth. Site 6 is located in front of the Bolinao town center. Site 5 is in a narrow water passageway south of Santiago Island. Sites 1, 2, 3 and 4 are outer reef sites. Sites 5 and 6 are inner sites. The Bolinao Marine Laboratory (BML: 16.38°N; 119.91°N) of the Marine Science Institute is also shown.

environment (see McManus et al. 1992). The bigger proportion of the marine (off-shore and near-shore) catch comes from the near-shore reef system. Seagrass meadows form the bigger part of the near-shore reef system (which includes the reef slope and reef flat), covering approximately 37 km^2 of the total 50 km^2 of reef flat area (Fortes 1989). From these meadows, fish and marine products (e.g. invertebrates and seaweeds) are harvested comprising ca. 77% of the total near-shore catch (McManus et al. 1992). For example, the top-ranking fish species by weight in shore landings in Bolinao (i.e., rabbitfish *Siganus fuscescens*, ca. 7 tons.$month^{-1}$) spends most of its lifetime in seagrass beds, i.e., it migrates only twice a year to the reef slopes for breeding (Aragones 1987). No doubt, seagrasses together with their associated flora and fauna form a substantial resource base for the economy of Bolinao.

To contribute to the understanding of the seagrass ecosystem in the area, data on various physico-chemical and biological factors affecting seagrass biology (in particular *Enhalus acoroides*) were collected during the period 1993-1996 at different sites (Fig. 2.1). From the initial visual survey, the criteria used for site selection were: (1) presence and absence of *Enhalus*; (2) visual differences in shoot size and density across a depth gradient; and (3) visual differences in water turbidity. In a collaborative research project on the hydrodynamics and sediment transport in the area, Rivera (1997) conducted a finer time-scale (weekly) collection of data on e.g. vertical light attenuation coefficient, temperature, salinity, sedimentation flux and water flow velocity, in 33 sampling sites around Santiago Island. The data on the above variables collected from the same eight study sites of that study were shared and are presented here in combination with the less frequent (once per 1-2 mos.) data collection in this study.

MATERIALS AND METHODS

Physico-chemical variables

For the data collection of vertical attenuation of photosynthetically active radiation (PAR, 400-700 nm), temperature, salinity, and water flow velocity, the same sampling strategy (site and frequency) was adopted, i.e., weekly from February 1994 - June 1995 at the sites. For sites 1a, 1b, 1c and 2, which were very close to each other, measurements were done only in one representative station (near 1c).

The PAR vertical attenuation coefficient of the water column was measured by taking simultaneous light readings at two depths (distance between light sensors > 0.5 m). The attenuation coefficient, K_d, was then calculated according to the function $I_z = I_o\ e^{-Kd*z}$, where I_z is the PAR recorded by the lower sensor, I_o is the PAR recorded by the upper sensor and z is the distance between sensors.

Water temperatures at the sites were measured using a thermometer connected to a digital display with 0.1 readability. Surface and bottom temperature differences did not exceed 0.5 °C. Data presented were obtained at bottom-depth.

Salinities were measured using a hand refractometer with 1‰ graduation. Readings were mostly done in the field stations or otherwise, water samples were brought to the Bolinao Marine Laboratory (BML, Fig. 2.1) in which case salinity was measured within the same day.

To measure sedimentation flux, sediment traps were installed at the sites. A sediment trap was made of a PVC pipe closed at one end (5 cm wide, 30 cm long; Rivera 1997). In the field stations, traps were fixed in a vertical position at the bottom and allowed to trap sediments for two weeks. During retrieval, contents of the traps were carefully transferred to properly labelled sampling bottles. Emptied traps were reinstalled for the next samples. At the laboratory, sediment samples were filtered using pre-dried and pre-weighed GF/C filters. The filtrates were then ovendried at 105 °C for 4 hours, cooled in a dessicator, and weighed. Values reported are in terms of $gDWm^{-2}.hr^{-1}$.

Water flow velocities were measured using the drogue method. A drogue was clocked while allowed to drift to a certain distance. Flow speed was then calculated as the distance travelled over the time elapsed and expressed in $cm.s^{-1}$. Flow directions were also recorded (cf. Rivera 1997) but are not presented in this report.

To characterize the sediment grain size distribution, three core samples (ca. top 10 cm) were collected from each site. Sediment samples were air-dried for four weeks to allow excess water to evaporate. Samples were then ovendried at 105°C to constant weights. Subsequently, sediments were sieved dry for 15 minutes using various pore sizes of screens stacked in a shaker. Before sieving, samples were gently crushed by hand to disintegrate well-sticked particles. Particles from site 5 samples were difficult to disintegrate but, nevertheless, sieved as above. After sieving, different size fractions were weighed and subsequently expressed as percentages of the total weight. Broader diameter size classes (in µm, e.g. coarse sand, 1000 > diameter > 250; fine sand, 250 > diameter > 125; very fine sand, 125 > diameter > 63; and silt, diameter < 63) were also computed by lumping appropriate specific fractions.

For sediment organic matter content, three sediment cores were collected from each site using a 20-ml syringe. Samples were ovendried at 105 °C to constant weights. Sediment samples were then combusted at 550 °C for four hours. After cooling (in a dessicator), ashes were weighed and the final values were expressed as % ash-free dry weights (AFDW).

Nutrients N and P in porewater and in the water-column were measured following spectrophotometric methods. However, for practical reasons, not all samples were analyzed at the same laboratory, depending on the cost of analyses and expertise available. Samples in May 1994 (water-column) were analized at the Chemical Oceanography Laboratory of the Marine Science Institute, University of the Philippines. Porewater samples in August and December 1994 were analized at the Chemical Engineering Laboratory of the College of Engineering, University of the Philippines. Samples in December 1995 were analyzed at the Chemical Laboratory of the International Institute for Hydraulic, Infrastructural and Environmental Engineering (IHE), Delft, The Netherlands (Knight 1996). Nutrients N, P, and K for sediments were determined after digestion at the Bureau of Soils and Water Management, Quezon City, Philippines. Soil samples in December 1995 were also analyzed at the Chemical Laboratory of IHE (Knight 1996).

Rainfall was measured at the Bolinao Marine Laboratory for the period January 1994 - December 1996. Using a rain gauge (a graduated cylinder, the height-volume factor of which had been calculated), rainfall was recorded as the total amount of rain accumulated over a day (24 hours; 0800H-0800H; expressed in $mm.d^{-1}$). Rivera (1997) collected the data for the period April 1994 - April 1995, afterwhich this study continued the collection till January 1996.

Tide levels were based on the predicted values published by the National Mapping and Resource Information Authority (NAMRIA, Philippines). Validations (Villanoy 1988; Rivera 1997) of these predictions revealed that only slight differences could be found between the predicted and measured values.

Daily total radiation data ($Joules.cm^{-2}d^{-1}$) were obtained from the National Radiation Center (Quezon City, Philippines) for the period 1992-1995. PAR values ($Einsteins.m^{-2}.d^{-1}$) were derived from these original values. Theoretical PAR values in a cloudless situation were also calculated. For details of the function formulation, see Philippart (1995).

Biological factors

The occurrence and approximate abundance of seagrasses and associated biota were noted throughout the study period. Density dynamics and recolonization data specific for *Enhalus acorides* (L.f.) Royle and *Thalassia hemprichii* (Ehrenb.) Aschers. were also collected and are presented in Chapter 4.

Mass accumulation and light attenuance (*sensu* Vermaat & Hootsmans 1994: proportional light reduction relative to a blank) by periphyton were studied in detail at sites 1a, 1b, 1c, 3, 5 and 6. Artificial substrates (transparent plastic strips, 80 x 20 mm) were incubated in the field for 28 days. Each substrate was properly tied to a small float and a sinker. In this way, substrates stood suspended and moved simulating *Enhalus* leaves. At day zero, eight substrates were thrown randomly within each of the sites. During retrieval, substrates were carefully transferred to labelled bags. Filtered seawater was added to each bag to prevent drying. After retrieval of all substrates from all sites, samples were transported to the laboratory. At the laboratory, light attenuance was determined. A light sensor was fixed inside a light-tight box with a small hole (diameter ~ 15 mm) on its topside. A light bulb was suspended directly above the box. Light readings with and without the periphyton substrates were taken. A blank strip had ca. 13% attenuance. Calculations were made to derive a one-side attenuance per strip (mean of 3 random points).

To measure biomass, the periphyton on each substrate was carefully scraped using the blunt edge of a scalpel. The periphyton was then filtered (pre-dried and pre-weighed GF/C filters) and ovendried at 60°C for 8 hours. After cooling, final periphyton weights were determined and expressed in $mgDWcm^{-2}$.

Mass accumulation and light attenuance were also studied in finer time scales at sites 1b and 5. In both sites at day zero, 72 units of artificial substrates were randomly thrown. Eight replicates were retrieved every five days. Further sample treatment procedures were done as above.

Data analyses

Significant differences between sites and between sampling months were tested using ANOVA following Sokal & Rohlf (1981). Whenever multiple comparisons were made, Tukey test was used (controlling the comparisonwise error rate, CER, to maintain an experimentwise error rate, EER, of 5%). When variance heterogeneity was detected, data were log-transformed which in all cases cured the problem.

RESULTS

Physico-chemical parameters

Light extinction and turbidity
The outer sites (1a, 1b, 1c, 2, 3 and 4) had consistently clear water with low light vertical attenuation coefficients (K_d) ranging from 0.1 to 0.7 m^{-1} (Fig. 2.2). The inner sites (5 and 6) had significantly ($p < 0.05$) higher mean K_d values and also wider variations (0.1-2.0 m^{-1}) than the outer sites. For any site, no significant differences could be found between months. For site 5, the relatively turbid water may be attributed to river runoff coming from the southern part of the area. Site 6 is located in front of an urban center, and may receive a considerable amount of wastewater especially from an adjacent fish market. Both inner sites (5 and 6) had an easily-resuspendable thin top layer of fine sediments.

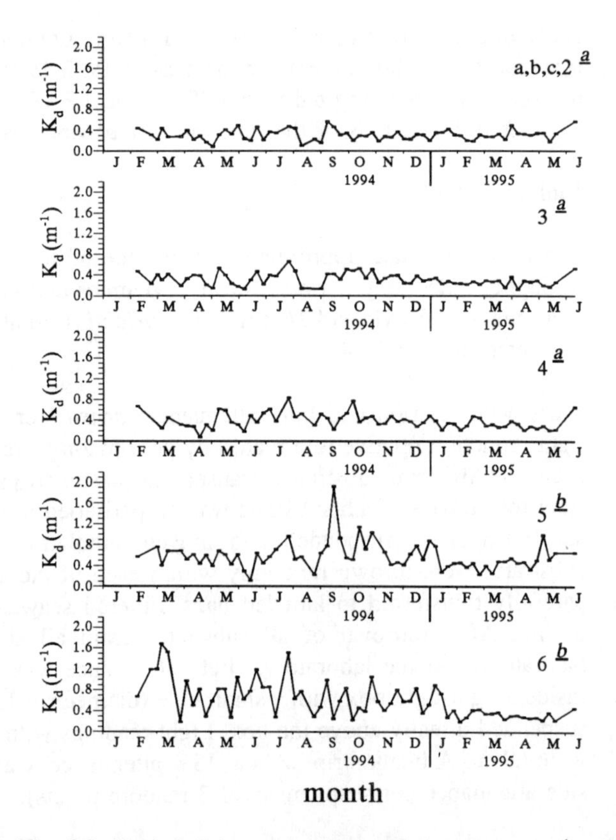

Fig. 2.2. Vertical light attenuation coefficient (K_d, in m^{-1}) measured at the different sites. Different underlined letters attached to site codes indicate significant differences ($p < 0.05$) between overall site means.

Temperature
Water temperature showed considerable variation over time, showing a bimodal trend (Fig. 2.3). No significant differences could be found between sites. Major peaks (32-35°C) were measured in April-May and September-October while major drops (26-28°C) were recorded in July-August and January-February. Temperature peaks and drops coincided between sites. The annual temperature range at any site throughout the study period was not more than 7 °C.

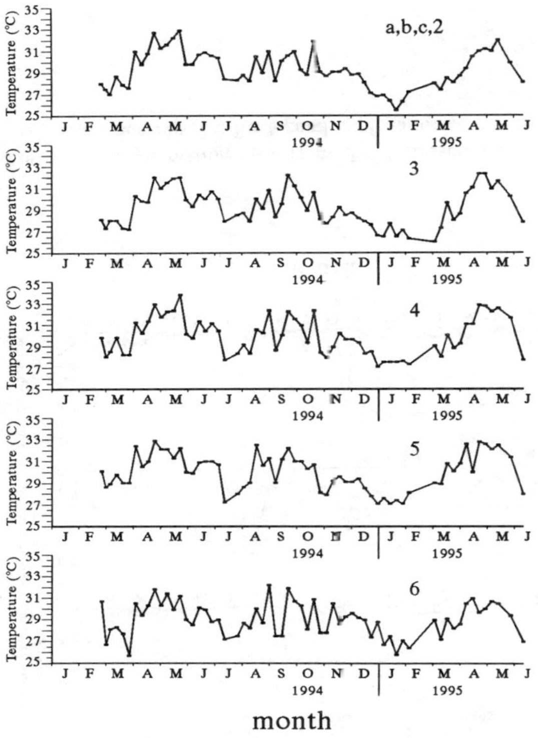

Fig. 2.3. Water temperature (°C) measured at the different sites. See Fig. 2.2 for legend explanations. No significant differences were found between any of the overal site means.

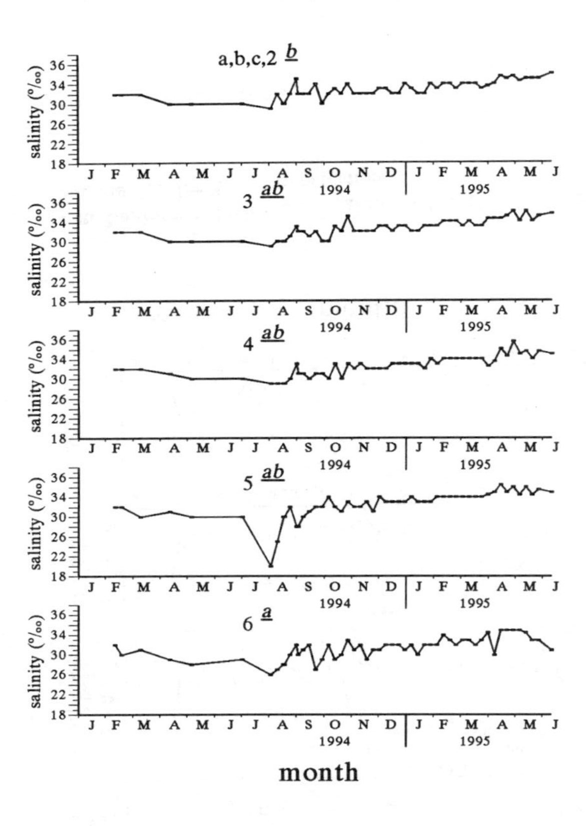

Fig. 2.4. Salinity (°/$_{oo}$) measured at the study sites. See Fig. 2.2 for legend explanations.

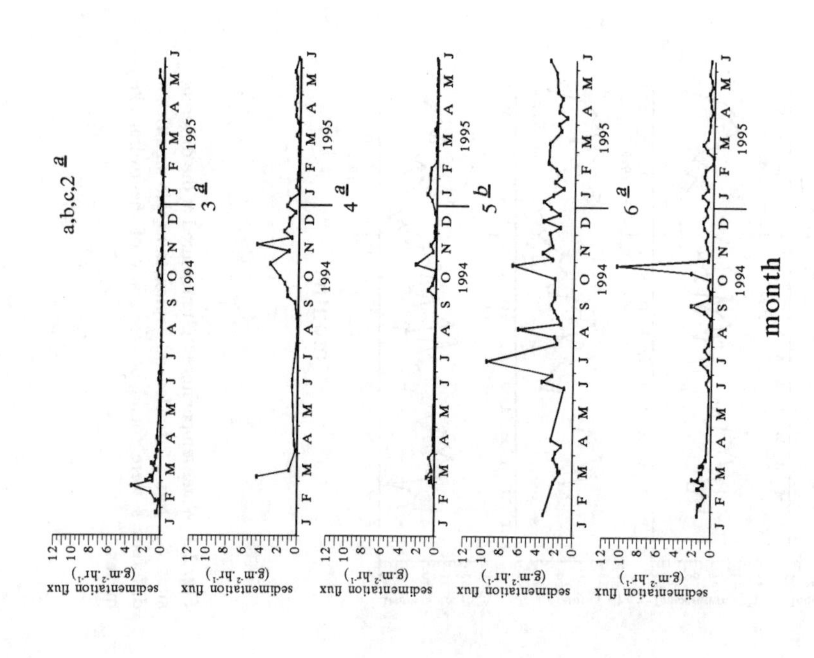

Fig. 2.5. Sedimentation flux ($g.m^{-2}.hr^{-1}$) measured at the study sites. See Fig. 2.2 for legend explanations.

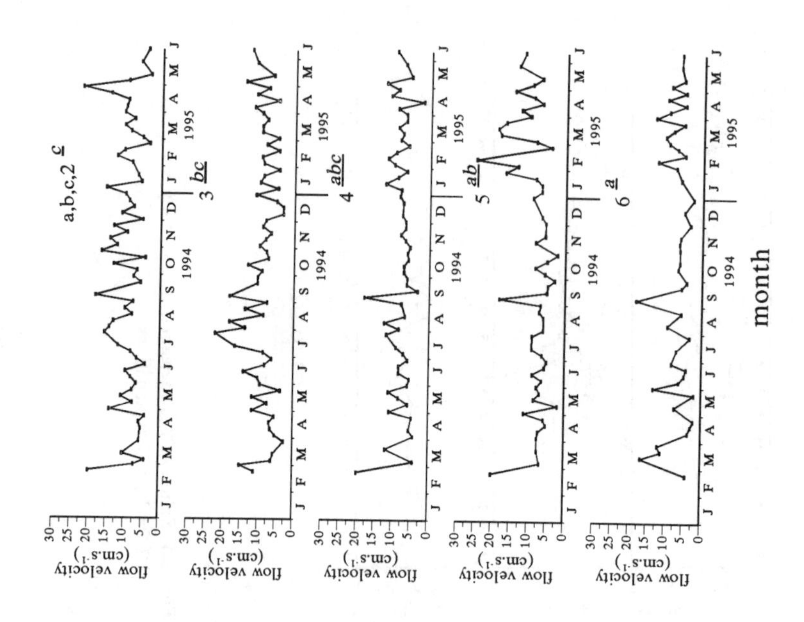

Fig. 2.6. Flow velocity ($cm.s^{-1}$) measured at the study sites. See Fig. 2.2 for legend explanations.

Salinity
No clear seasonality in salinity could be detected (Fig. 2.4). Values obtained for sites 1, 2, 3 and fluctuated between 30-35$^{o}/_{oo}$. Site 5 had comparable values with an exceptional drop (20$^{o}/_{oo}$) in early August 1994 (shortly after few days of continous rainfall). The lower mean salinity and wider variation in site 6, most probably, was due to the high amounts of wastewater from the dense human settlement fronting this site.

Sedimentation flux
There was no substantial sedimentation occurring in most of the sites (1,2 3, 4 and 6) with values less than 1.5 gDW $m^{-2}.hr^{-1}$ (Fig 2.5). Still, however, small peaks could be observed (October-November 1994; February-March 1995), probably an aftermath of a stormy period (June-September 1994) or sustained winds (mid-January to early March 1995; see also Rivera 1997). The effect of stormy conditions on the sedimentation flux was especially apparent in site 3, a deep site (5 m below zero datum) with only occasionally, sparse shoots of *Halophila ovalis* (pers. obs.). Sedimentation in site 5 was significantly higher than in all other sites (Fig. 2.5; averaging 3.73 ± 0.31 gDW $m^{-2}.hr^{-1}$). Sedimentation in site 5 was also more dynamic, that is, consequences of related factors e.g., rains, storms and sustained winds, could be immediately recognizable. For instance, high values (the highest approached 10 gDW $m^{-2}.hr^{-1}$, mid-July 1994) were recorded coinciding with stormy days and continuous rainfall (see also Figs. 2.4 and 2.8). The almost instantaneous effects of other factors on the sedimentation dynamics in site 5 could be attributed to the facts that: (1) sediment substrate in this site is loosely-packed and is highly resuspendable with only little disturbance; (2) average water depth is below 1 m and; (3) there is strong influence of riverine waters coming from a bigger bay south of the study area.

Water flow velocity
Comparing sitewise, sites 1, 2 and 3 had relatively stronger flow velocities, averaging > 9 $cm.s^{-1}$ (Fig. 2.6) while sites 5 and 6 had less flow with mean values of 6.7 ± 0.76 $cm.s^{-1}$ and 7.2 ± 0.54 $cm.s^{-1}$, respectively. Overall, differences were only slight (see also Rivera 1997) i.e., only a few centimeters per second. For all sites, the data did not show strong seasonality.

Sediment grain-size distribution
The sediment grain-size profiles of the sites (Table 2.1) indicated that all were mainly composed of medium to very coarse sand (~60-75% of oven-dry weight, DW), fine sand (~20-30% DW), very fine sand (~5-20% DW) and silt (ranging from a neglible amount to ~5% DW). Remarkable differences could be observed between sites 2 and 3 and all the other sites because of their higher composition of the medium-very coarse sand (> 70%) fraction as well as the lower amount of the very fine sand (only ~5%). Site 5 and 6 did not show much difference from other sites but it should be noted that in these sites the topsoil is composed of a very thin layer of highly resuspendable silt. Also, the color characteristics of the sediments in sites 5 and 6 (from greyish to black) were strikingly in contrast with those from all other sites (yellowish to white).

Sediment organic matter content
The organic matter contents of the sediments from the outer reef sites were all similar (~5% DW, Fig. 2.7). The inner sites 5 and 6 had significantly higher organic matter content (8-9% DW).

Table 2.1. Summary of the sediment grain-size composition data. Values are in %DW ± se, n = 3.

	SITES							
	1a	1b	1c	2	3	4	5	6
(grain size, μm)								
> 850	14.12 ± 1.20	19.62 ± 1.51	21.64 ± 1.32	24.85 ± 0.44	17.33 ± 1.77	14.31 ± 0.15	11.75 ± 1.47	15.53 ± 1.34
> 600 to 850	16.36 ± 0.73	17.04 ± 0.76	18.80 ± 0.81	23.45 ± 0.90	16.24 ± 1.29	14.20 ± 0.58	12.44 ± 0.45	16.10 ± 0.66
> 425 to 600	15.48 ± 0.48	11.82 ± 0.70	12.55 ± 0.51	14.86 ± 0.76	14.53 ± 1.09	13.24 ± 0.64	13.82 ± 0.46	14.69 ± 0.67
> 355 to 425	8.40 ± 0.38	5.72 ± 0.08	5.75 ± 0.24	6.01 ± 0.30	8.63 ± 0.44	7.67 ± 0.15	9.17 ± 0.16	9.78 ± 0.10
> 250 to 355	10.89 ± 0.95	6.30 ± 0.10	6.57 ± 0.79	7.30 ± 0.55	15.82 ± 4.88	8.13 ± 0.27	12.63 ± 0.58	12.12 ± 0.92
> 180 to 250	11.26 ± 0.65	8.87 ± 0.91	11.31 ± 0.69	10.36 ± 0.65	13.84 ± 2.02	14.88 ± 1.53	18.51 ± 0.82	13.11 ± 0.84
> 125 to 180	8.85 ± 0.59	9.18 ± 1.26	10.53 ± 0.62	7.26 ± 1.17	7.06 ± 0.72	9.64 ± 0.65	10.70 ± 1.32	8.87 ± 1.87
> 106 to 125	6.52 ± 0.92	11.68 ± 1.97	9.07 ± 1.21	2.79 ± 0.22	2.71 ± 0.70	8.75 ± 1.47	4.65 ± 0.78	4.03 ± 0.57
> 63 to 106	4.25 ± 1.46	9.32 ± 0.75	3.40 ± 0.61	2.78 ± 0.36	3.51 ± 0.47	6.75 ± 2.53	5.34 ± 0.34	4.71 ± 0.43
< 63	3.86 ± 2.51	0.45 ± 0.11	0.40 ± 0.23	0.33 ± 0.12	0.33 ± 0.07	2.42 ± 1.61	0.99 ± 0.36	1.07 ± 0.20
WENTWORTH SCALE								
medium-very coarse sand (> 250 μm)	65.25 ± 0.96	60.50 ± 2.89	65.29 ± 2.51	76.48 ± 2.10	72.55 ± 2.76	57.55 ± 1.00	59.81 ± 1.79	68.21 ± 2.41
fine sand (> 125 to 250 μm)	20.12 ± 0.67	18.05 ± 0.60	21.84 ± 1.31	17.62 ± 1.80	20.90 ± 2.73	24.53 ± 2.00	29.21 ± 1.18	21.98 ± 2.69
very fine (> 63 to 125 μm)	10.77 ± 1.89	21.00 ± 2.53	12.47 ± 1.10	5.57 ± 0.29	6.22 ± 0.24	15.50 ± 2.62	9.99 ± 0.84	8.74 ± 0.84
silt and clay (< 63 μm)	3.86 ± 2.51	0.45 ± 0.11	0.40 ± 0.23	0.33 ± 0.12	0.33 ± 0.07	2.42 ± 1.61	0.99 ± 0.36	1.07 ± 0.20

Nutrients

water-column

Nitrogen concentrations in the water-column showed wide variation and to infer meaningful comparison is difficult (Table 2.2) . For instance in May 1994, ammonium concentration (which comprise > 80% of the total dissolved inorganic nitrogen) at S1a was found significantly higher ($p < 0.05$) than either at S1b and S1c. S1b was only 15 m (horizontal distance) away from S1a while S1c was 50 m away. With site 3 also having low concentration (see 1994 and all the values in December 1995; Table 2.2), these results suggest an effect of water depth. However, corresponding concentrations at site 2 and site 4 in 1994 do not fit into this correlation. Similar observations could be made for phosphorus (Table 2.2).

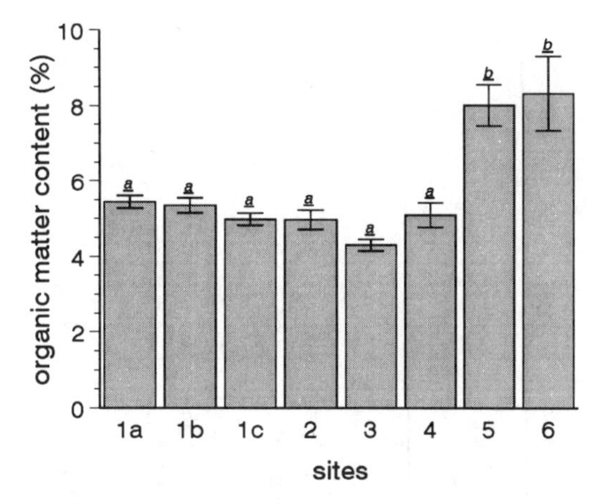

Fig. 2.7. Organic matter content (%) of sediment samples from the study sites. Different underlined letters above bars indicate significant ($p < 0.05$) differences between the means.

porewater

Relative to the concentrations in the water-column, porewater concentrations of nitrogen were higher (Table 2.2; 16 - 86 μM with ammonium comprising 70% of total N). Sites 2 and 3 tended to have lower concentrations. No clear difference was shown between inner (sites 5 and 6) and outer (sites 1, 2 ,3 and 4) sites. Phosphorus concentrations in the porewater were also higher than the corresponding water-column values. No differences between sites could be found.

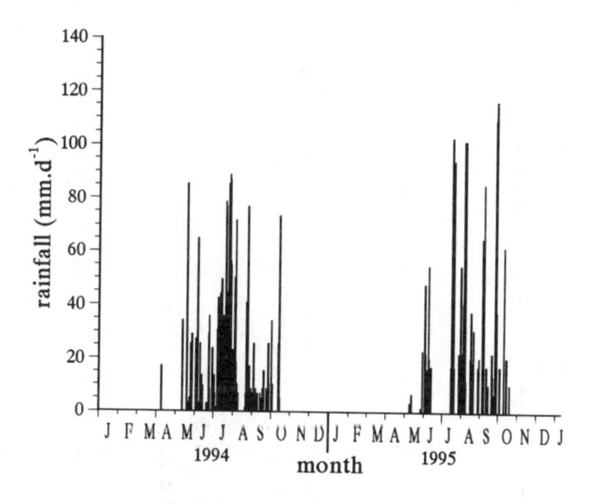

Fig. 2.8. Rainfall ($mm.d^{-1}$) measured at the Marine Science Institute-Bolinao Marine Laboratory (BML, see Fig. 2.1) for the period February 1994-December 1995.

sediment

Sediments from inner sites had higher nitrogen and potassium contents than those from outer reef sites (Table 2.3). For phosphorus content and pH, no distinct differences could be found.

Rainfall

Distinct dry and rainy seasons were observed for Bolinao during the period January 1994 to January 1996. The rainy period started about early May and ended mid-October (Fig. 2.8).

Table 2.2. Nitrogen and phosphorus concentrations (± se, n = 3) in the water column and porewater.

	SITES							
	1a	1b	1c	2	3	4	5	6
Water column								
NH_4-N (μM)								
May 1994	24.85 ± 9.75	4.88 ± 1.78	4.54 ± 0.69	9.34 ± 1.81	5.48 ± 2.01	14.24 ± 2.56	25.71 ± 5.45	9.11 ± 2.49
Dec 1995	18.9 ± 1.8	17.4 ± 3.2	8.2 ± 1.0	10.0 ± 0.0	1.7 ± 0.5	8.9 ± 1.1	18.6 ± 1.9	14.2 ± 1.5
NO_3-N (μM)								
May 1994	2.33 ± 0.21	0.61 ± 0.26	0.78 ± 0.25	1.16 ± 0.33	0.60 ± 0.14	1.68 ± 0.46	6.23 ± 2.83	1.85 ± 0.18
Dec 1995	1.2 ± 0.2	0.7 ± 0.3	0.7 ± 0.2	0.6 ± 0.4	0.0 ± 0.0	7.6 ± 0.6	8.4 ± 0.1	11.3 ± 0.8
NO_2-N (μM)								
May 1994	0.91 ± 0.09	0.40 ± 0.06	0.38 ± 0.03	0.56 ± 0.05	0.31 ± 0.05	0.68 ± 0.09	0.96 ± 0.19	0.87 ± 0.09
PO_4-P (μM)								
May 1994	6.94 ± 0.51	2.67 ± 1.50	2.16 ± 0.70	4.54 ± 0.48	2.09 ± 0.95	4.67 ± 0.84	8.23 ± 1.11	8.65 ± 0.31
Dec 1995	0.23 ± 0.01	0.28 ± 0.03	0.17 ± 0.01	0.05 ± 0.01	0.22 ± 0.00	0.39 ± 0.01	0.52 ± 0.02	0.15 ± 0.03
Porewater								
NH_4-N (μM)								
Aug 1994	72.57 ± 4.31	79.39 ± 8.15	75.58 ± 3.04	44.36 ± 0.98	42.41 ± 2.93	47.32 ± 0.40	86.19 ± 1.35	67.04 ± 5.50
Dec 1994	45.76 ± 4.35	57.00 ± 2.05	58.43 ± 0.62	45.55 ± 2.57	46.33 ± 0.99	68.53 ± 5.74	65.79 ± 6.32	58.69 ± 5.35
Dec 1995	67.7 ± 4.3	64.8 ± 4.8	53.0 ± 5.6	80.3 ± 1.0	15.9 ± 0.2	45.0 ± 0.7	51.9 ± 0.5	70.5 ± 3.4
NO_3-N (μM)								
Dec 1995	8.8 ± 0.3	5.1 ± 2.3	7.7 ± 0.1	6.7 ± 0.9	9.2 ± 0.2	7.3 ± 0.6	8.6 ± 1.6	11.8 ± 2.3
PO_4-P (μM)								
Dec 1995	3.88 ± 0.73	3.16 ± 0.07	2.26 ± 0.18	2.68 ± 0.08	2.86 ± 0.30	1.93 ± 0.17	2.22 ± 0.22	2.15 ± 0.33

Table 2.3. Nutrient contents and pH of sediments (± se; n = 3).

	SITES							
	1a	1b	1c	2	3	4	5	6
total N (%)								
Jan 1995	0.093 ± 0.019	0.073 ± 0.003	0.053 ± 0.003	0.037 ± 0.003	0.050 ± 0.000	0.087 ± 0.012	0.127 ± 0.007	0.153 ± 0.012
Jun 1995	0.068 ± 0.004		0.058 ± 0.003	0.045 ± 0.004	0.052 ± 0.002		0.087 ± 0.007	0.155 ± 0.006
Dec 1995	0.175 ± 0.066	0.174 ± 0.009	0.071 ± 0.024	0.080 ± 0.026	0.056 ± 0.009	0.175 ± 0.013	0.214 ± 0.080	0.354 ± 0.021
total P (ppm)								
Mar 1994	15.7 ± 0.9	17.0 ± 0.6	11.3 ± 0.7	11.0 ± 1.2	20.7 ± 1.5	14.3 ± 0.3	19.7 ± 2.9	30.7 ± 2.6
Jul 1994	11.3 ± 0.9	11.7 ± 0.3	8.7 ± 1.8	10.0 ± 1.0	17.0 ± 0.6	11.0 ± 0.6	18.0 ± 3.0	12.3 ± 1.9
Jan 1995	19.3 ± 3.0	13.3 ± 2.4	11.3 ± 1.2	9.0 ± 0.6	19.3 ± 1.5	14.3 ± 1.2	14.0 ± 1.5	23.3 ± 0.9
Jun 1995	7.6 ± 0.4		11.6 ± 0.4	6.6 ± 0.4	16.2 ± 0.8		8.8 ± 0.7	11.7 ± 0.3
Dec 1995, %	0.021 ± 0.001	0.019 ± 0.001	0.024 ± 0.000	0.032 ± 0.004	0.027 ± 0.001	0.027 ± 0.002	0.024 ± 0.002	0.028 ± 0.002
total K (ppm)								
Mar 1994	346.7 ± 20.3	295.0 ± 2.9	191.7 ± 15.9	195.0 ± 17.6	203.3 ± 17.6	293.3 ± 3.3	841.7 ± 92.8	575.0 ± 14.4
Jul 1994	225.3 ± 47.2	190.7 ± 31.5	161.3 ± 7.4	185.3 ± 4.8	170.7 ± 9.3	262.0 ± 68.2	616.0 ± 129.3	502.7 ± 143.5
Jan 1995	321.7 ± 58.3	238.3 ± 14.2	151.7 ± 7.3	151.7 ± 6.0	153.3 ± 17.4	336.7 ± 72.2	941.7 ± 141.7	476.7 ± 61.7
Jun 1995	205.6 ± 5.8		193.1 ± 5.4	158.8 ± 5.3	169.3 ± 7.7		574.6 ± 44.6	412.1 ± 7.1
pH								
Jul 1994	8.33 ± 0.17	8.43 ± 0.07	8.67 ± 0.03	8.77 ± 0.03	8.37 ± 0.13	8.20 ± 0.15	8.23 ± 0.03	8.37 ± 0.03
Jun 1995	8.72 ± 0.03		8.73 ± 0.03	9.10 ± 0.03	9.00 ± 0.03		8.65 ± 0.04	8.52 ± 0.03

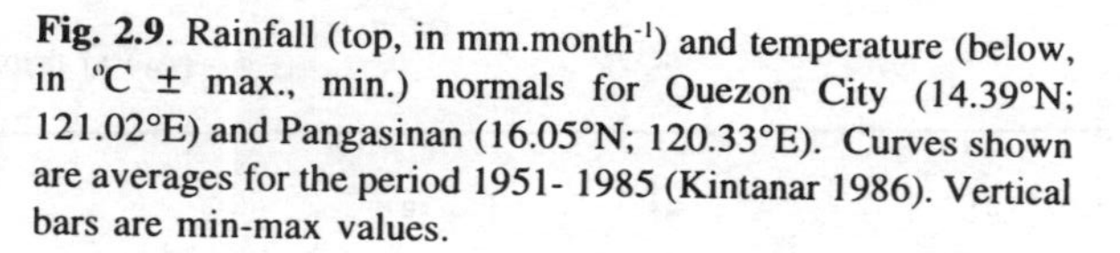

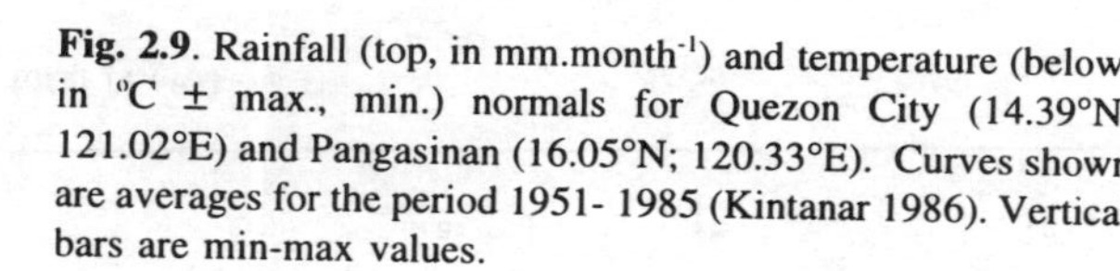

Fig. 2.9. Rainfall (top, in mm.month^{-1}) and temperature (below, in °C ± max., min.) normals for Quezon City (14.39°N; 121.02°E) and Pangasinan (16.05°N; 120.33°E). Curves shown are averages for the period 1951- 1985 (Kintanar 1986). Vertical bars are min-max values.

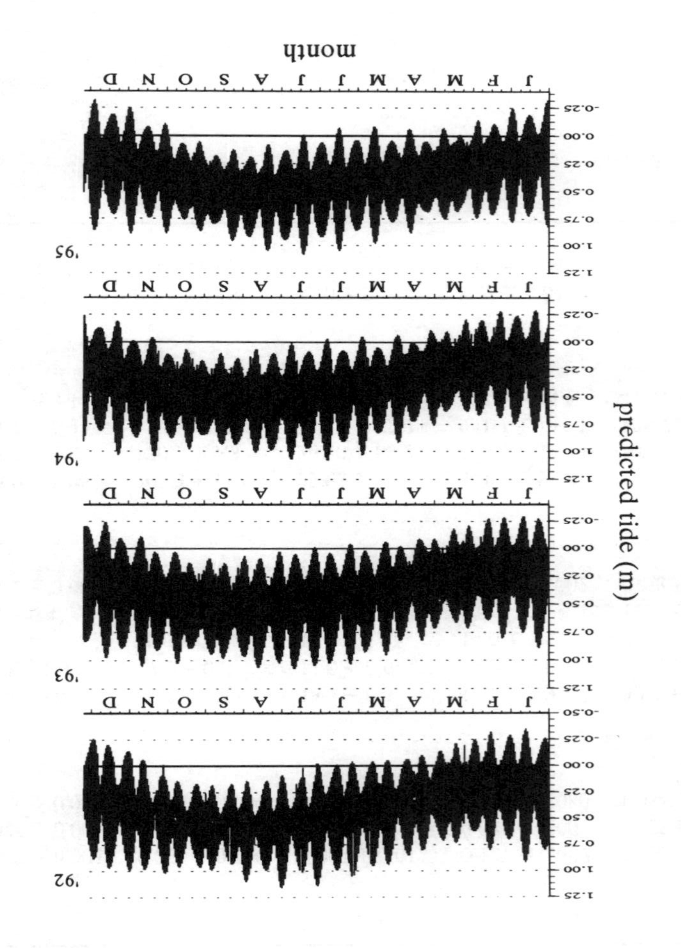

Fig. 2.10. Tide level profiles (in meters) for the years 1992, 1993, 1994 and 1995 (data source: tide tables, National Mapping and Resource Information Authority, NAMRIA, Philippines).

Rainfall peaks were around July-August resulting in short-term drops in salinity at especially site 5 and, less strongly, site 6 (see also Fig. 2.4). This seasonal trend in rainfall agrees very well with the rainfall normals for Pangasinan province for the period 1951-1985 (Fig. 2.9).

Tides

Tidal fluctuation in Bolinao for the years 1992, 1993, 1994 and 1995 (Fig. 2.10) was semi-diurnal with varying amplitude (max. of 1.3 m during spring tides; minimum of 0.2 m during neap tides). The time intervals between spring and neap tides were similar over the years. However, the mean water level differed between months, although the tidal ranges between months were similar. For these years, the mean water level was high in the June-September period and low during December to February with a difference of about 0.4 m.

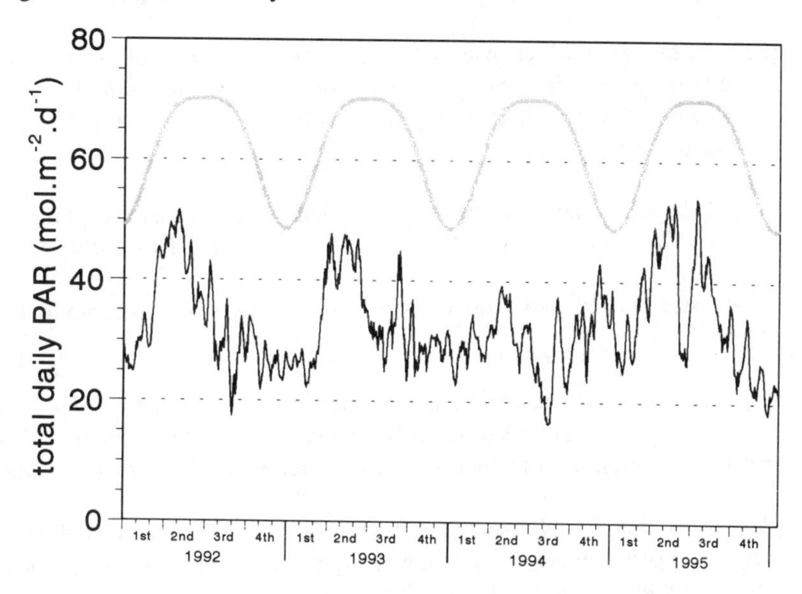

Fig. 2.11. Total daily radiation (PAR, in Einsteins.$m^{-2}.d^{-1}$). Observed values (15-day moving average, curve below) were derived from the data (in Joules.$cm^{-2}.d^{-1}$) collected at a nearest weather station (National Radiation Center, Quezon City: 14.39°N; 121.02°E) recording global insolation. Theoretical PAR (thick line; smooth curve) represents a clear-sky situation (for derivations, see Philippart 1995).

Global radiation

The total daily radiation recorded at the earth's surface (14.39°N, 121.02°E) and the corresponding theoretical radiation in a clear-sky condition are shown in Fig. 2.11. Without clouds (smooth curve), the amount of radiation distinctly varies within a year, that is, peaks in June-July (ca. 70 Einsteins.$m^{-2}.d^{-1}$) and lows in December-January (ca. 50 Einsteins.$m^{-2}.d^{-1}$). Data recorded at the weather station (unstable curve) indicated substantial PAR reductions by clouds averaging ca. 50%. Highest reductions were recorded during rainy seasons (July-August 1993-1995; see also Fig. 2.8). This global radiation pattern is supposed to be also prevailing in Bolinao (16.38°N, 119.91°E) as supported by the temperature and rainfall normals for the two areas (1951-1985 data; Fig. 2.9; Kintanar 1986).

Biological factors

community structure overview

site 1: Many species of seagrasses coexisted in this site. The number of species decreased as water depth increased. The shallow portion (1a) was a mixed bed of *Enhalus acoroides, Thalassia hemprichii, Halodule uninervis* (wide and narrow varieties), *Cymodocea rotundata, Cymodocea serrulata, Syringodium isoetifolium* and *Halophila ovalis*. At 1b, *Thalassia* and *Enhalus* stood taller and densities of other species became less. At 1c, the tallest stands of *Enhalus* and *Thalassia* were found with occasional occurrence of *Halodule* and *Halophila.* At 1c, the density of *Cymodocea serrulata* was highest. At this depth (3.0 m below zero datum), no shoots of *Syringodium isoetifolium* and *Cymodocea rotundata* could be found. Further, seasonal abundance of the seaweed *Halimeda macroloba* was noted.

site2: In this area which was only a few meters away from site 1c, no seagrass meadow existed except the infrequent growth of *Halophila* and *Halodule*. In this area, the particularly high density of Thalassinid shrimp mounds (i.e., *Callianasa* sp., ca. 4-5 m^{-2}) was remarkable compared to site 1a.

site 3: Like site 2, there was no seagrass meadow in this area except the intermittent incidence of *Halophila*. No occurrence of *Halodule* was observed. This site was 5 m below zero datum.

site 4: Here existed also a mixed seagrass bed closely comparable to 1b in water depth and vegetation.

site 5: Here existed a relatively patchy meadow dominated by *Enhalus acoroides, Thalassia hemprichii* and tall shoots of *Halodule uninervis*. During December - February period every year (1993-1996), high incidence of green urchins (*Salmacis sphaeroides*) was observed.

site 6: This area was characterized by a thick forest of *Enhalus* and *Thalassia* with no other seagrass existing under the canopy. Frequent algal scum could be observed on the water surface. Often, *Enhalus* leaves were epiphytized with filamentous green macroalgae (*Chaetomorpha* sp.). Jellyfish (*Cassiopea* sp.) was persistent here with peak abundance during March-May.

Mass accumulation and light attenuance by periphyton

Mean periphyton biomass accumulation on artificial leaves per 28-day incubation ranged from ~1 $mgDW.cm^{-2}$ to ~2.5 $mgDW.cm^{-2}$. This range corresponded into light attenuance ranging from a low 35% to a high 75% (Fig. 2.12). Comparing between sites, higher values in both periphyton biomass accummulation and light attenuance were found at sites 3 and 5. Sites 1a and 1b had lower values in both parameters. In general, biomass curves showed a unimodal trend (peak in September of 1994; low in January), a pattern more or less followed by the corresponding attenuance curves (Figs. 2.13 and 2.14). Periphyton characteristics differed between sites as could be inferred from their attenuance-biomass curves (Fig. 2.15). For instance, at site 6, a biomass of 11 $mgDW.cm^{-2}$ only corresponded to ~80% attenuance while much lower biomass for sites 3 and 5 (e.g., 3 $mgDW.cm^{-2}$) corresponded to ~90% light reduction. Periphyton biomass at site 6 could be high but was mainly composed of green algae (qualitative data, in prep) while at sites 3 and 5, silt contents were higher. Sites 1a, 1b and 1c had fairly similar periphyton quality. These qualitative and quantitative differences could further be shown by comparing, on a finer time scale, periphyton colonization and light attenuance between sites 1b and 5 (Fig. 2.16). Biomass accumulation was approximately linear for both sites with the slope for site 5 being significantly steeper. Comparing the attenuance-biomass ratios, the difference was also clear, e.g., a similar biomass of 1.5 $mgDW.cm^{-2}$ translated to different attenuance values of ca. 50% and 80% for sites 1b and 5, respectively.

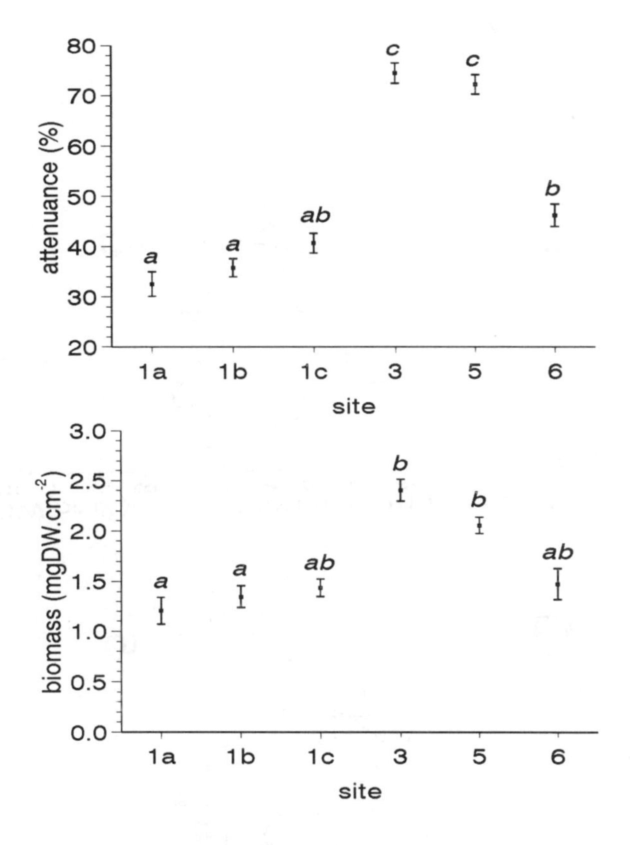

Fig. 2.12. Mean periphyton light attenuance (top) and biomass accumulation (bottom) studied in 6 sites. Error bars are standard errors (n = 8). Different letters in italics show significant differences ($p < 0.05$) between site means.

DISCUSSION

Across sites, some of the physico-chemical characteristics differed, although the range was generally narrow in magnitude. Clearly, the inner sites had more turbid waters (Fig. 2.2), more organic matter (Fig. 2.7), and more nitrogen (Table 2.3) and potassium (Table 2.3) contents in the sediments than the outer sites. Site 5 (an inner site) also had a significantly higher sedimentation flux (Fig. 2.5). Of the outer sites, sediment particles in sites 2 and 3 had both high percentages of the medium to very coarse sand and less of the very fine sand (Table 2.2). No other abiotic parameter was found to be distinctly different across outer sites.

The magnitude of the differences in light extinction coefficients across sites was not enormous. A mean difference of about 0.3 m^{-1} (site 1a vs 5: 0.3 ± 0.02 vs. 0.6 ± 0.4 m^{-1}) was not much compared to that of the carbonate (0.34 ± 0.13 m^{-1}) vs. terrigenous sites (1.59 ± 0.75 m^{-1}) in Indonesia (Erftemeijer 1993). More turbid waters over seagrass beds can be found in other studies (Philippart 1995, Terschelling, The Netherlands, K_d ranging from 2 to 4 m^{-1}; Zimmerman et al. 1995, Paradise Cove, California, K_d range: 1 - 2.5 m^{-1}; Van Lent et al. 1991, Mauritania, mean K_d = 1.61 ± 0.10 m^{-1}; Vermaat & Verhagen 1996, Zandkreek estuary, SW Netherlands, K_d = 2.1 ± 0.1 m^{-1}).

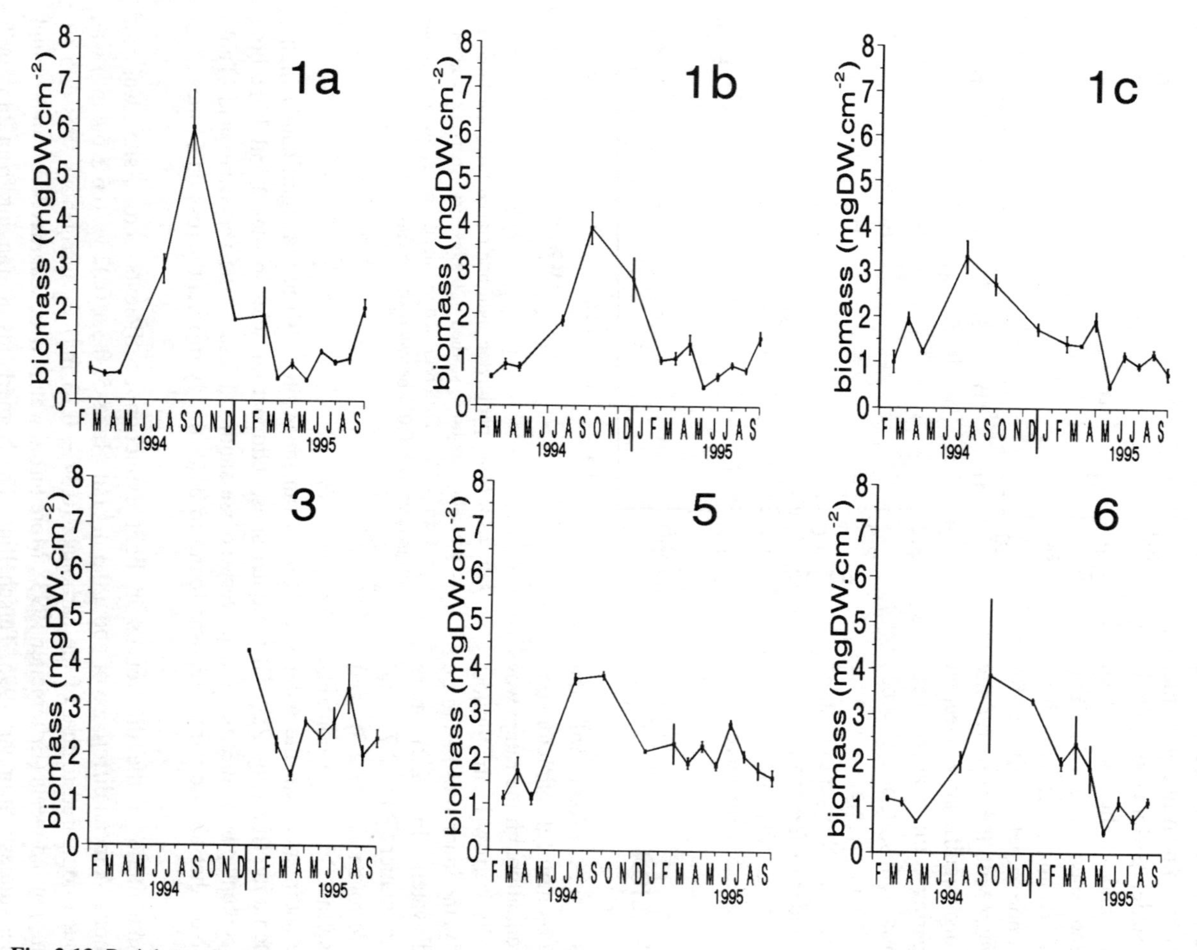

Fig. 2.13. Periphyton biomass accumulation (28-day period) . Site codes are indicated in the upper-left side of each graph. Error bars are standard errors, n = 8.

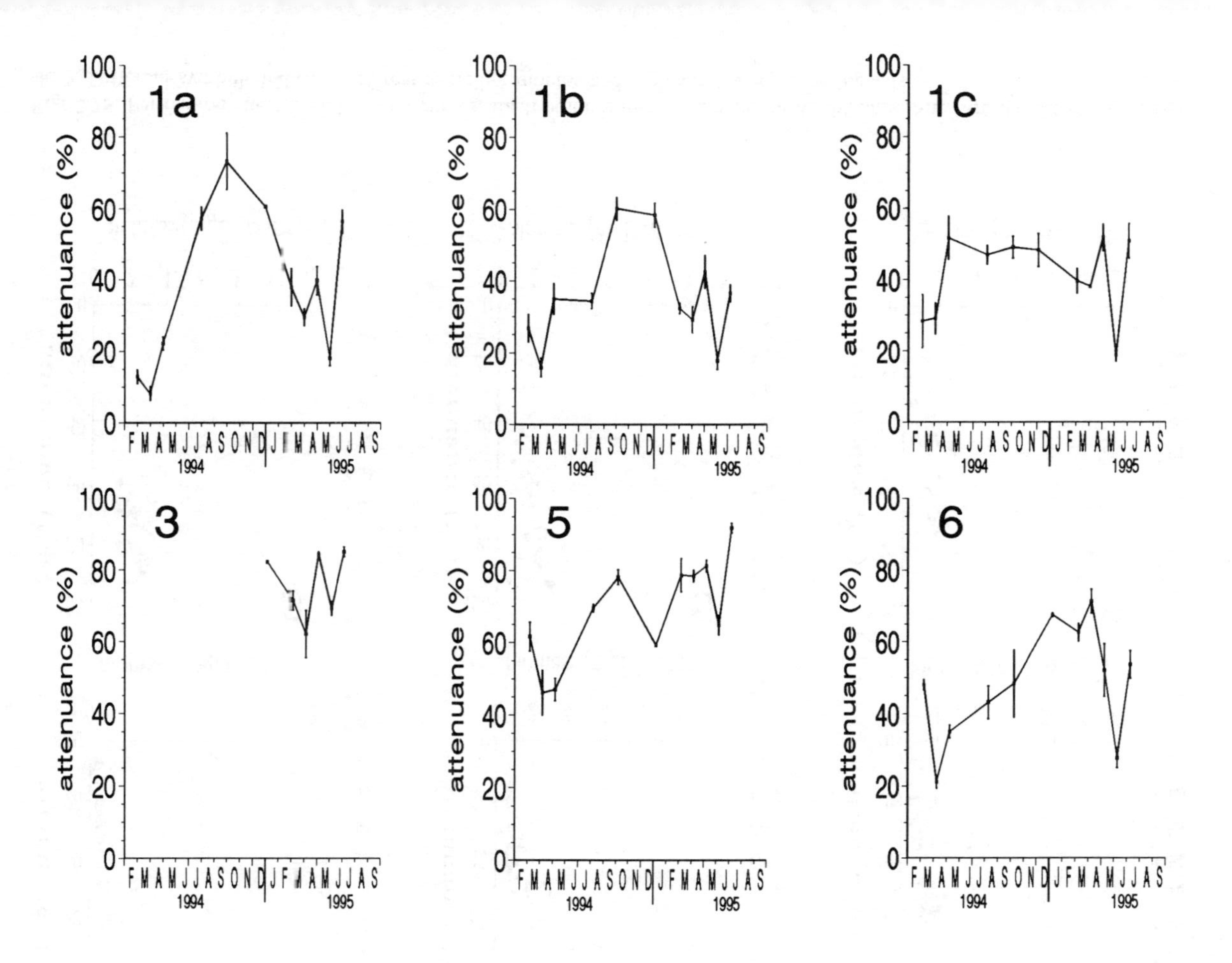

Fig. 2.14 Periphyton light attenuance. See Fig. 2.13 for legend details.

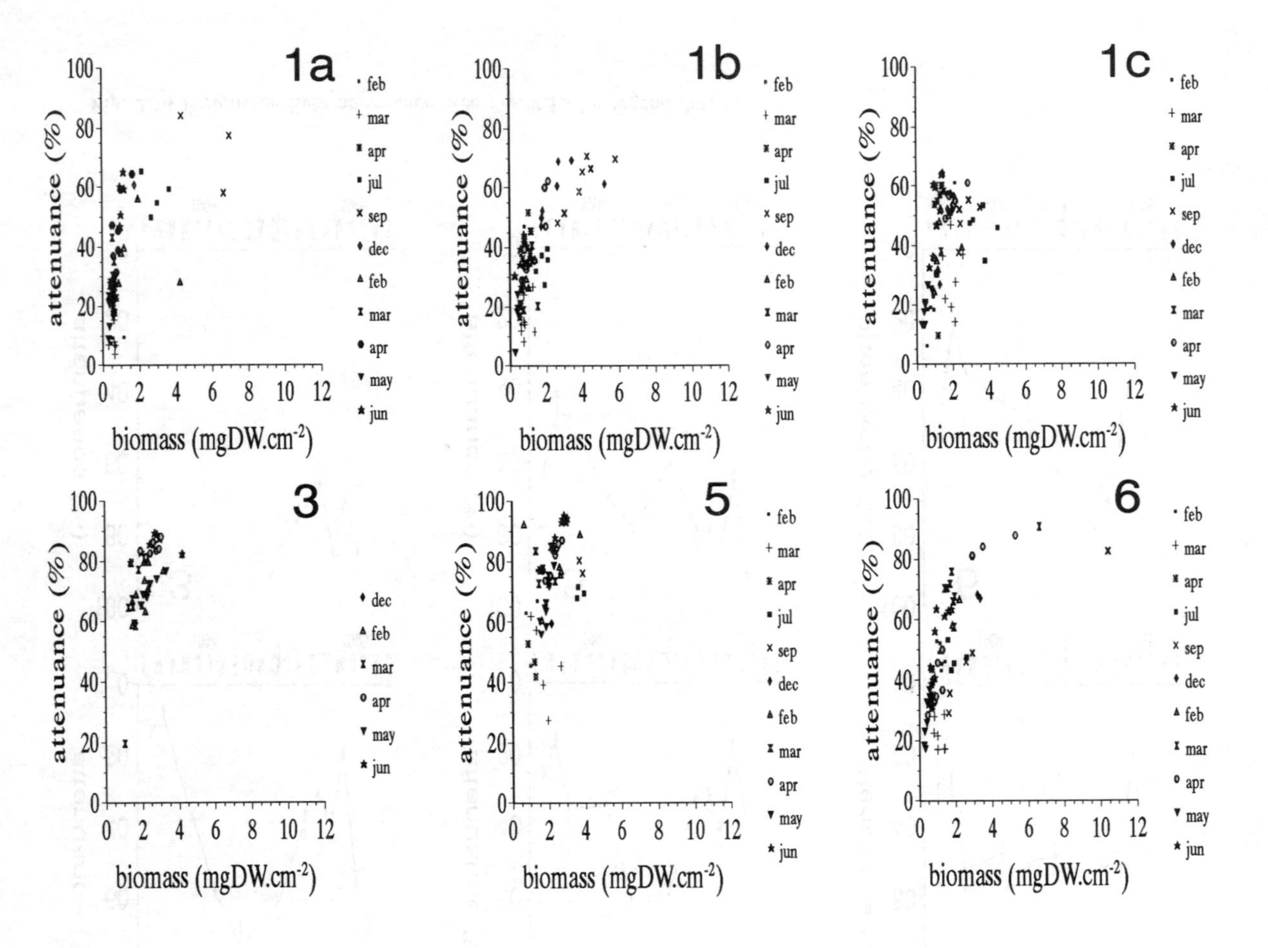

Fig. 2.15. Periphyton attenuance-biomass curves. Each point represents a mean of 8 replicates. Sites are indicated (upper-right side). Different symbols indicate different sampling months and are enumerated in the legends.

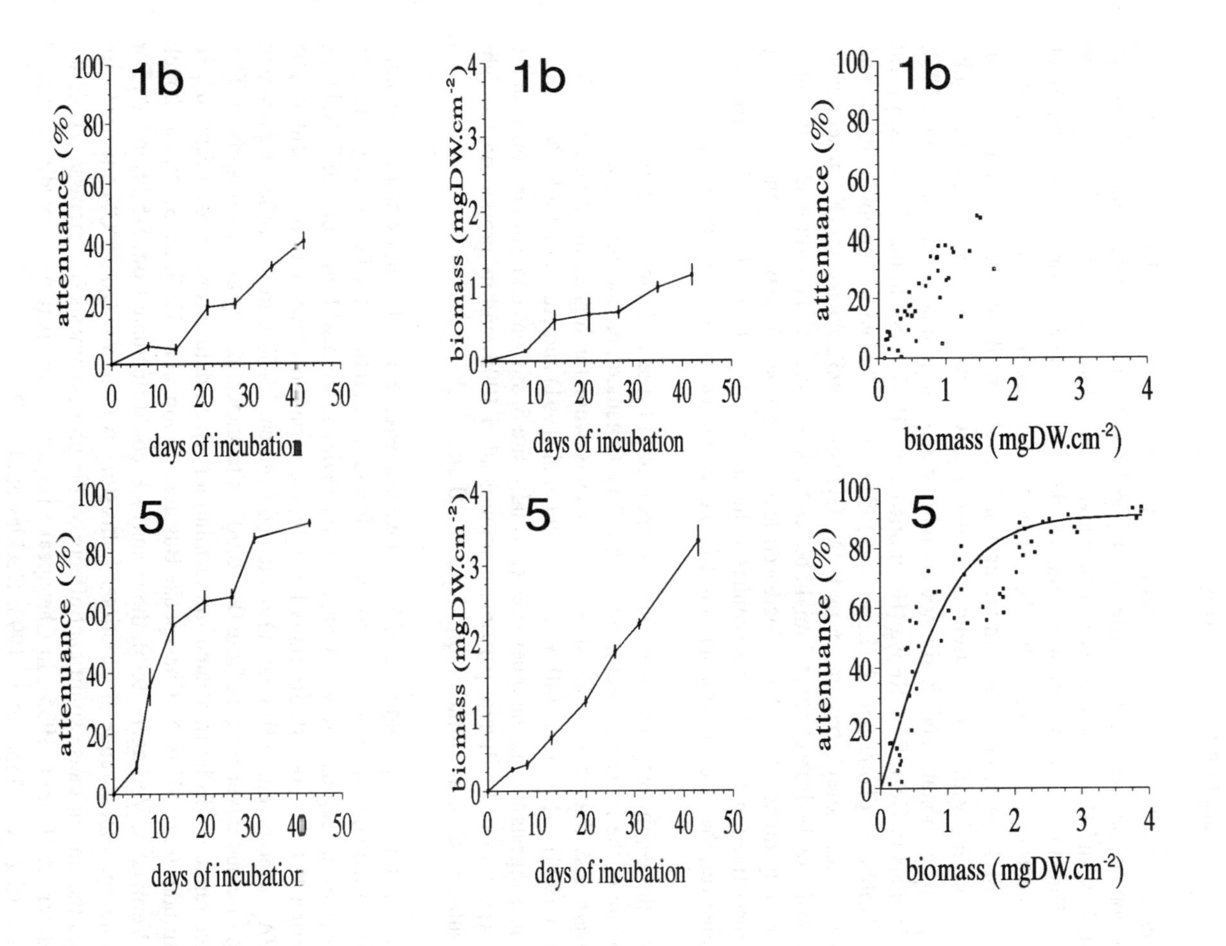

Fig. 2.16. Finer time scale (high frequency) comparison of colonization and light attenuance between sites 1b and 5. Error bars are standard errors, n = 8. Attenuance-biomass graphs (third column) show all samples.

Salinity may have effects on seagrasses (e.g. Koch & Dawes 1991). However, the salinity of waters with connection to open oceans would not vary too widely (e.g. Banc d' Arguin, Mauritania, Dedah 1993; Sulawesi, Indonesia, Erftemeijer 1993). Results from this study showed an intermittent influence of freshwater at the inner sites, although in general, the magnitude of the variation was weak (Fig. 2.4).

The concentrations of nutrients in the water column were clearly lower than in the porewater. The same conclusion was made by Erftemeijer (1993) in comparable environments. However across sites, N and P concentrations lacked convincing differences. This result may not be surprising. Due to its highly dynamic nature, definite conclusions on the differences of nutrient concentrations are difficult when these differences are not, at least, 10-fold in magnitude (Gil Jacinto, personal communication, 1993). Small events, e.g. rain and tidal flushing, could easily disrupt the trends. Similar observations can be made on the results comparing the carbonate and terrigenous environments in Indonesia (Erftemeijer 1993), different habitats in Gazi, Bay, Kenya (Hemminga et al. 1995), sites in the Florida Keys and western Caribbean (Tomasko & Lapointe 1991), and between eutrophic inner and mesotrophic outer Bailey's bay in Bermuda (McGlathery 1995). Potassium (K) in sediments showed considerable differences and might better classify sites, i.e. the sediments at the inner sites had clearly higher K contents. Increased levels of potassium may, in general, directly affect carbon fixation (rate of photosynthesis) in higher plants (Marschner 1986; Ericsson 1995). However, the effect of this nutrient (K) on seagrasses has not been explicitly studied.

Results on the depth-integrated grain size composition (Table 2.1) did not show the 'silty' characteristic of the inner site sediments (Table 2.1). Obtained percentages of particle size < 63 µm were less than 5% in all sites. Silt contents were far higher in Indonesia (40%, Erftemeijer 1993) , Mauritania (40-60%, van Lent et al. 1991) and Australia (40%, McKenzie 1994). This suggests that at the inner sites especially site 5, silt was concentrated only on the top layer (pers. obs.; P. Rivera, personal communication 1996) and its proportional weight was not substantial in a sediment-core depth of 10 cm.

Over time, the total daily irradiance (PAR) above sea level (Fig. 2.11) and water temperature (Fig. 2.3) varied significantly. Variation in PAR and temperature may be causally linked to the differences in seagrass growth (Chapter 3; for reviews, see also Dennison 1987, Bulthuis 1987, Vermaat et al. in press). Specially for light, the variation shown here is substantial. The 15-day PAR average curve (Fig. 2.11) fluctuated from ca. 15-50 $E.m^{-2}.d^{-1}$ reflecting a strong influence of cloud cover ranging from 0% to 90% attenuance (daily values, not shown). This attenuance range could be translated into a minimum-maximum figure of 5-70 $E.m^{-2}.d^{-1}$. In higher latitudes (e.g., 37°16'N, Chesapeake Bay area, Moore et al. 1993; 51°35'N, Zandkreek estuary, Vermaat & Verhagen 1996), the mean PAR could fluctuate from 5-50 $E.m^{-2}.d^{-1}$. For temperature, the annual range of variations in this study was narrow (26-35°C) compared with the ranges that might cause significant effects on the photosynthesis of seagrasses (Bulthuis 1987; Barbers & Behrens 1985). In Chesapeake Bay, water temperature annually fluctuated much more (5-25°C, Moore et al. 1993; Neckles et al. 1993).

Despite the overall lack of large differences in physico-chemical parameters, the biological environment considerably varied between sites. The communities differed in composition and structure, ranging from a completely mixed but 'dwarfed' meadow at site 1a (a total of seven seagrass species together with several species of benthic macroalgae, e.g. *Halimeda* sp.) to

Table 2.4. Deterioration of light climate conditions at the different sites. $PAR_{theoretical}$ = annual mean for the study area (Figs. 2.1 and 2.11); cloud/rain factor = ratio of $PAR_{weather\ station}$ and $PAR_{theoretical}$; water surface reflection ~10% (Scheffer et al. 1993); k-depth factor = transmittance due to the extinction coefficient and mean water depth; mean water depth = site depth with respect to datum chart plus the mean tidal fluctuation; periphyton figures are mean transmittance percentages (Fig. 2.12); * = assumed value, no data; $PAR_{seagrass\ level}$ = remaining PAR at seagrass depth; %$PAR_{theoretical}$ = percentage of PAR value with respect to $PAR_{theoretical}$; % PAR = remaining PAR at seagrass depth with respect to $PAR_{sea\ level}$ (incident light).

	SITES							
	1a	1b	1c	2	3	4	5	6
$PAR_{theoretical}$, in $E.m^{-2}.d^{-1}$ (cloudless condition)	60	60	60	60	60	60	60	60
Transmittance through:								
cloud/rain	52.9%	52.9%	52.9%	52.9%	52.9%	52.9%	52.9%	52.9%
surface reflection	90.0%	90.0%	90.0%	90.0%	90.0%	90.0%	90.0%	90.0%
k-depth factor	76.8%	65.9%	42.8%	41.6%	23.0%	37.4%	42.7%	62.1%
periphyton	67.5%	64.3%	59.4%	60.0%*	25.6%	60.0%*	27.9%	53.8%
$PAR_{seagrass\ level}$, in $E.m^{-2}.d^{-1}$ (periphyton excluded)	21.93	18.81	12.29	11.83	6.64	18.81	12.17	17.65
$PAR_{seagrass\ level}$, in $E.m^{-2}.d^{-1}$ (periphyton included)	14.8	12.1	7.3	7.1	1.7	6.4	3.4	9.5

Table 2.5. Summary of conclusions in comparing the means of different environmental parameters across sites. Site-independent variables, e.g. rainfall, insolation, are not included; different letters indicate significant differences at $p < 0.05$ (ANOVA, Tukey test), i.e., a < b < c etc.; ns, not significant; ***, test not possible; **, if significant difference exists; max. difference, maximum difference (highest minus lowest) between site means; mixed, all 7 seagrass species were present; ET, *Enhalus-Thalassia*; ETC, ET-*Cymodocea serrulata*; ETH, ET-*Halodule*; bare, no seagrasses

	SITE								
parameter	1a	1b	1c	2	3	4	5	6	max. difference
physico-chemical									
1. vertical light attenuation coefficient, k_d	a	a	a	a	a	a	b	b	0.3 m^{-1}
2. temperature	ns	ns	ns	ns	ns	ns	ns	ns	
3. salinity	b	b	b	b	ab	ab	ab	a	$2^{0}/_{00}$
4. sedimentation flux	a	a	a	a	a	a	b	a	2.75 $gDW.m^{-2}.h^{-1}$
5. water flow velocity	c	c	c	c	bc	abc	ab	a	4 $cm.s^{-1}$
6. sediment grain size									
medium-coarse sand	a	a	a	b	b	a	a	a	15%
very fine sand	b	b	b	a	a	b	b	b	10%
other fractions	ns	ns	ns	ns	ns	ns	ns	ns	
7. sediment organic matter content	a	a	a	a	a	a	b	b	4%
8. depth with respect to datum chart	a	b	c	d	e	b	a	a	4.5 m
9. PAR at depth (periphyton included)	14.8	12.1	7.3	7.1	1.7	6.4	3.4	9.5	13 $E.m^{-2}.d^{-1}$
10. PAR at depth (periphyton excluded)	21.9	18.8	12.3	11.8	6.6	18.8	12.2	17.7	15.3 $E.m^{-2}.d^{-1}$
11. nutrients									
N water-column	ab	a	a	ab	a	ab	b	ab	20 µM
porewater	ns	ns	ns	ns	ns	ns	ns	ns	
sediment	a	a	a	a	a	a	b	b	0.1%
P water-column	bc	a	a	abc	a	abc	c	c	7 µM
porewater***	-	-	-	-	-	-	-	-	2 µM**
sediment	ab	ab	a	a	bc	ab	bc	c	10 ppm
K sediment	a	a	a	a	a	a	c	b	600 ppm
biological									
12. community structure	mixed	mixed	ETC	bare	bare	mixed	ETH	ET	
13. periphyton									
light attenuance	a	a	ab		c		c	b	42%
biomass	a	a	ab		b		b	ab	1.2 $mgDW.cm^{-2}$
fine-scale rates		low					high		

a thick and tall forest of only *Enhalus acoroides* and/or *Thalassia hemprichii* at sites 5 and 6. Also, colonization and light attenuance by periphyton across sites were different (Figs. 2.12 to 2.16). Periphyton could negatively affect seagrass photosynthesis (Sand-Jensen 1977; Bulthuis & Woelkerling 1983; Silberstein et al. 1986; Neckles et al. 1994) but in certain cases, grazers could reduce the effect of periphyton (Hootsmans et al. 1993; Neckles et al. 1994; Philippart 1995). Light attenuance by periphyton has been estimated elsewhere (max. ~25% of incident light, Hootsmans et al. 1993; 10-90%, Philippart 1995; 9%, Vermaat & Verhagen 1996). This study obtained comparable values and further showed that periphyton attenuance in specific sites could translate into substantial light reduction (Table 2.4). Assuming that the values obtained from artificial subtrates approximate the colonization on the natural leaves, then the residual PAR at site 3 (Table 2.4) would be below the survival threshold for *Enhalus acoroides* (see also Vermaat et al. 1995 for required LCP for photosynthesis; Chapter 8).

CONCLUSIONS

1.) Table 2.5 summarizes the major conclusions from spatial comparisons of the various variables measured. In short, inner sites differed in particular from outer sites in terms of vertical light attenuation coefficient, sedimentation rates, sediment organic matter, total sediment N & K contents and periphyton characteristics.

2.) Several variables, namely incident light, rainfall and temperature distinctly showed seasonal variation. Salinity might be fairly constant over a year but, in some areas as in site 5, could drop substantially over short periods after freshwater influx.

In view of the above, the succeeding chapters (3-8) will focus on the influence of light and nutrients on growth and reproduction of *Enhalus acoroides*.

LITERATURE CITED

Aragones, N.V. 1987. Taxonomy, distribution, and relative abundance of juvenile siganids and aspects of the *padas* fishery in Bolinao, Pangasinan, Philippines. Marine Science Institute, University of the Philippines, Diliman, Quezon City, Philippines. MS Thesis, 66 pp.

Barber, B.J. and J.P. Behrens. 1985. Effects of elevated temperature on seasonal *in situ* leaf productivity of *Thalassia testudinum* Banks ex König and *Syringodium filiforme* Kützing. Aquat. Bot. 22: 61-69.

Bulthuis, D.A. 1987. Effects of temperature on photosynthesis and growth of seagrasses. Aquat. Bot. 27: 27-40.

Bulthuis, D.A. and W.J. Woelkerling. 1983. Biomass accumulation and shading effects of epiphytes on the leaves of *Heterozostera tasmanica*, in Victoria, Australia. Aquat. Bot. 16: 137-148.

Dedah, S.O. 1993. Wind, surface water temperature, surface salinity and pollution in the area of the Banc d'Arguin, Mauritania. Hydrobiologia 258: 9-19.

Dennison, W.C. 1987. Effects of light on seagrass photosynthesis, growth and depth distribution. Aquat. Bot., 27: 15-26.

Ericsson, T. 1995. Growth and shoot:root ratio of seedlings in relation to nutrient availability. Plant and Soil 168-169: 205-214.

Erftemeijer, P.L.A. 1993. Factors limiting growth and production of tropical seagrasses: nutrient dynamics in Indonesian seagrass beds. PhD thesis, Katholieke Universiteit Nijmegen, The Netherlands, 173 pp.

Fortes, M.D. 1989. Seagrasses, a resource unknown in the ASEAN region. ICLARM Education Series 5, International Centre for Living Aquatic Resources Management, Manila.

Hemminga, M.A., P. Gwada, F.J. Slim, P. de Koeyer and J. Kazungu. 1995. Leaf production and nutrient contents of the seagrass *Thalassodendron ciliatum* in the proximity of a mangrove forest (Gazi Bay, Kenya). Aquat. Bot. 50: 159-170.

Hootsmans, M.J.M, J.E. Vermaat and J.A.J. Beijer. 1993. Periphyton density and shading in relation to tidal depth and fiddler crab activity in intertidal seagrass beds of the Banc d'Arguin (Mauritania). Hydrobiologia, 258: 73-80.

Kintanar, J. 1986. Philippine climatological normals. PAG-ASA, The Philippines.

Knight, D.L.E. 1996. Allocation of nutrients in the seagrass *Enhalus acoroides*. MSc thesis, IHE, Delft, The Netherlands, 34 pp.

Koch, E.W. and C.J. Dawes. 1991. Ecotypic differentiation in populations of *Ruppia maritima* L. germinated from seeds and cultured under algal-free laboratory conditions. J. Exp. Mar. Biol. Ecol., 152: 145-159.

Marschner, H. 1986. Mineral nutrition of higher plants. Academic Press, London, 674 pp.

McGlathery, K.J. 1995. Nutrient grazing influences on a subtropical seagrass community. Mar. Ecol. Prog. Ser. 122: 239-252.

McKenzie, L.J. 1994. Seasonal changes in biomass and shoot characteristics of a *Zostera capricorni* Aschers. dominant meadow in Cairns Harbour, North Queensland. Aust. J. Mar. Freshwater Res. 45: 1337-1352.

McManus, J.W., C.L. Nañola, Jr., R.B. Reyes, Jr. and K.N. Keshner. 1992. Resource ecology of the Bolinao coral reef system. ICLARM Stud. Rev. 22, 117 pp.

McManus, L.T. and T.E. Chua. 1990. The coastal environmental profile of Lingayen Gulf, Philippines. ICLARM Technical Reports 22, 69 pp.

Moore, K.A., R.J. Orth and J.F. Nowak. 1993. Environmental regulation of seed germination in *Zostera marina* L. (eelgrass) in Chesapeake Bay: effects of light, oxygen and sediment burial. Aquat. Bot. 45: 79-91.

Neckles, H.A., R.L. Wetzel and R.J. Orth. 1993. Relative effects of nutrient enrichment and grazing on epiphyte-macrophyte (*Zostera marina* L.) dynamics. Oecologia 93: 285-295.

Neckles, H.A., E.T. Koepfler, L.H. Haas, R.L. Wetzel and R.J. Orth. 1994. Dynamics of epiphytic photoautotrophs and heterotrophs in *Zostera marina* (eelgrass) microcosms: responses to nutrient enrichment and grazing. Estuaries 17: 597-605.

Philippart, C.J.M. 1995. Seasonal variation in growth and biomass of an intertidal *Zostera noltii* stand in the Dutch Wadden Sea. Neth. J. Sea. Res. 33: 205-218.

Rivera, P.C. 1997. Hydrodynamics, sediment transport and light extinction off Cape Bolinao, Philippines. PhD dissertation, IHE-WAU, The Netherlands, 244 pp.

Sand-Jensen, K. 1977. Effects of epiphytes on eelgrass photosynthesis. Aquat. Bot. 3: 55-63.

Scheffer, M., A.H. Bakema and F.G. Wortelboer. 1993. MEGAPLANT: a simulation model of the dynamics of submerged plants. Aquat. Bot. 45: 341-356.

Silberstein, K., A.W. Chiffings and A. McComb. 1986. The loss of seagrass in Cockburn Sound, Western Australia. III. The effects of epiphytes on productivity of *Posidonia australis* Hook. f. Aquat. Bot. 24: 355-371.

Sokal, R. and F.J Rohlf. 1981. Biometry. The principles and practice of statistics in biological research, 2nd ed. WH Freeman and Co., New York, 859 pp.

Tomasko, D.A. and B.E. Lapointe. 1991. Productivity and biomass of *Thalassia testudinum* as related to water column nutrient availability and epiphyte levels: field observations and experimental studies. Mar. Ecol. Prog. Ser. 75: 9-17.

Van Lent, F., P.H. Nienhuis and J.M. Verschuure.1991. Production and biomass of the seagrass *Zostera noltii* Hornem. and *Cymodocea nodosa* (Ucria) Aschers. at the Banc d'Arguin (Mauritania, NW Africa): a preliminary approach. Aquat. Bot. 41: 353-367.

Vermaat, J.E and M.J.M. Hootsmans. 1994. Periphyton dynamics in a temperature-light gradient. In: W. van Vierssen, M.J.M. Hootsmans and J.E. Vermaat (eds.). Lake Veluwe, a macrophyte-dominated system under eutrophication stress. Geobotany 21. Kluwer, Netherlands, 193-212 pp.

Vermaat, J.E. and F.C.A. Verhagen. 1996. Seasonal variation in the intertidal seagrass *Zostera noltii* Hornem.: coupling demographic and physiological patterns. Aquat. Bot. 52: 259-281.

Vermaat, J.E., C.M Duarte and M.D. Fortes. 1995. Latitudinal variation in life history patterns and survival mechanisms in selected seagrass species, as a basis for EIA in coastal marine ecosystems (Final Report). Project EC DG XII-G CI1*-CT91-0952, IHE Delft, The Netherlands, 38 pp. + 2 annexes.

Vermaat, J.E., N.S.R. Agawin, C.M. Duarte, S. Enriquez, M.D. Fortes, N. Marba, J.S. Uri and W. van Vierssen. In press. The capacity of seagrasses to survive eutrophication and siltation, across-regional comparisons. Ambio.

Villanoy, C. 1988. Inferred circulation of Bolinao coral reef. Technical Report, Marine Science Institue, University of the Philippines, Diliman, Quezon City, The Philippines.

Zimmerman, R.C., J.L. Reguzzoni and R.S. Alberte. 1995. Eelgrass (*Zostera marina* L.) transplants in San Francisco Bay: Role of light availability on metabolism, growth and survival. Aquat. Bot. 51: 67-86.

Chapter 3

Spatio-temporal variation in shoot size and leaf growth of the two dominant Philippine seagrasses *Enhalus acoroides* (L.f.) Royle and *Thalassia hemprichii* (Ehrenb.) Aschers.

Abstract. Variation in shoot size and leaf growth of the two dominant Philippine seagrasses *Enhalus acoroides* and *Thalassia hemprichii* was investigated over a 2-year period at different sites. To evaluate the influence of light on this variation, calculations of the daily photosynthetic rates were done using a photosynthesis-irradiance curve (Michaelis-Menten model), and site-specific light climate estimates. Light as PAR was appropriately corrected for cloud cover attenuation, water surface reflectance, tides, water depth and water turbidity. Subsequently, the daily net photosynthetic rates (= net production) were compared with the corresponding relative growth rates in the field to quantify how much of (1) the net production is allocated to leaf growth and (2) the seasonal variation in leaf growth is due to seasonality in net photosynthesis, i.e., as a function of PAR.

Results showed that for both species, shoot sizes are larger in deeper and darker environments. The leaf relative growth rates (RGR) across sites remain comparable except for the slightly higher RGR values of *E. acoroides* in muddy sites. The influence of light on the seasonality in RGR was shown to be species- and site-specific. The seasonal effect of light on *E. acoroides* was generally significant explaining ca. 25-43% of the seasonal variation in leaf growth. However, in a shallow- and clear-water site, light seasonality showed little influence (r^2 = 6%; p = 0.42) on the RGR of *E. acoroides*. For *T. hemprichii*, the influence of light on the seasonality in RGR accounted for less than 2% of the variation in all sites.

Further decomposing the partial contributions of the different components in the photosynthesis model showed that at least 89% of the variation in photosynthesis was due to the cloud cover attenuation component. Expressing the measured relative growth rate as a percentage of net photosynthesis, plants in deeper and darker environments allocate more photosynthate to leaf growth (for *E. acoroides* ca. 100% in the deepest site vs. < 85% in shallower and clear waters; for *T. hemprichii*, the greatest allocation was exhibited by plants growing in the deepest site). This suggests that plants in deeper and darker environments probably will have less photosynthate available for other functions, e.g., flower, root and rhizome production. For *Enhalus*, this is supported by observed lower flowering frequencies.

INTRODUCTION

In the tropics, seagrasses form extensive meadows that perform several important trophic and structural functions. In the Indo-West Pacific region, seagrass beds are multi-species (Johnstone 1979; Meñez et al. 1983; Brouns 1985, 1987a, 1987b; Erftemeijer 1993; Tomasko et al. 1993; Vermaat et al. 1995a) often dominated by *Enhalus acoroides* (L.f.) Royle and *Thalassia hemprichii* (Ehrenb.) Aschers. (Vermaat et al. 1995a). In Bolinao, NW Philippines, these two species together comprise ca. 90% of multi-species meadow biomass (Vermaat et al 1995a). Thus, understanding the biology of these two species is necessary to substantially gain insights into the behaviour of seagrass meadows in the area.

For leaf growth of *E. acoroides* and *T. hemprichii*, some substantive studies have been done (mainly by Johnstone 1979, Brouns 1985, 1987 and Brouns & Heijs 1986 for Papua New Guinea; Erftemeijer 1993 for Indonesia; Estacion & Fortes 1988, Tomasko et al. 1993 and

Vermaat et al. 1995a for the Philippines). Although in these studies, spatial variability in growth could be attributed to the environmental characteristics of the corresponding site, seasonal patterns lack concrete explanations. For *Enhalus acoroides,* a seasonal, unimodal pattern was observed in Papua New Guinea (Brouns & Heijs 1986) while a similar study in the Philippines showed a bimodal trend (Estacion & Fortes 1988). Brouns and Heijs (1986) explained seasonality by the strong positive correlation of growth with temperature data. Estacion and Fortes (1988) were more circumstantial in explaining growth variability as a complex function of daylength and the occurrence of low, low water (LLW) during the day. In that study, growth correlated positively with daylength and negatively with LLW but also, growth peaks occurred during warm months. For *Thalassia hemprichii*, studies on growth seasonality are even more scarce. Brouns (1985) found no consistent trend on a year-to-year basis, i.e., a high value was observed in June 1982 but the lowest value was recorded in June 1981. This contrast has been attributed to tidal influence. Tidal exposure and water movement have also been shown to have primary influence on growth characteristics of *Enhalus acoroides* and *Thalassia hemprichii* in Sulawesi, Indonesia (Erftemeijer 1993).

So far, factors proposed to explain seasonal trends of growth in *Enhalus acoroides* and *Thalassia hemprichii* have been more correlative than causal. A more direct factor such as available irradiance has not yet been explicitly incorporated. For instance, the correlations of growth with water level (Brouns 1985; Estacion & Fortes 1988) and daylength (Estacion & Fortes 1988) might be refined if changes in photosynthetically active radiation (PAR) due to these factors are determined. Indeed, temperature may have a direct effect on growth (Barber & Behrens 1985; Bulthuis 1987) but seasonality in temperature is also a reflection of the mean trend in insolation.

In this chapter, the *in situ* dynamics of leaf growth in *Enhalus acoroides* and *Thalassia hemprichii* in Bolinao, NW Philippines are investigated in relation to factors which affect available PAR. The approach was: (1) to measure *in situ* several growth parameters of plants at different sites with different light characteristics (cf. Chapter 2) over a 2-year period; (2) to predict corresponding (time and space) photosynthetic rates as a function of PAR, appropriately corrected for cloud cover attenuation, water surface reflection, tide level, water depth and water turbidity; and (3) to relate the outcome of these two.

MATERIALS AND METHODS

In situ measurements

Leaf growth of *Enhalus acoroides* and *Thalassia hemprichii* was studied in sites 1a, 1b, 1c, 4, 5 and 6 (see Fig. 2.1, Chapter 2) from June 1993 to February 1996 for *E. acoroides* and from February 1994 to September 1995 for *T. hemprichii.* During each sampling, 20 shoots of both species were randomly selected. Each shoot was marked with colored, thin wires, and tiny holes were punched through all the leaves using a hypodermic needle (Zieman 1974; modified marking technique). The sheaths of the outermost non-growing leaves were used as reference levels. After about 15 days, all the marked shoots were harvested and the following variables were measured in the laboratory for each shoot:

a.) AGR, leaf areal growth rate ($cm^2.sht^{-1}.d^{-1}$)

$$AGR = \frac{\Sigma \; leaf \; length \; increments \; * \; width}{observation \; period, \; days}$$

b.) TSA, one-sided total leaf surface area ($cm^2.sht^{-1}$)

$$TSA = \Sigma \; leaf \; lengths \; * \; widths$$

c.) RGR, relative growth rate ($cm^2.cm^{-2}.sht^{-1}.d^{-1}$; the fraction of leaf material produced per unit time; van Lent et al. 1995)

$$RGR = \frac{AGR}{TSA}$$

Photosynthesis calculations

Model inputs

Using a photosynthesis-irradiance curve (Michaelis-Menten model; for review of other models see Jassby & Platt 1976; Frenette et al. 1993; Hootsmans & Vermaat 1994; Santamaria-Galdon & van Vierssen 1995), the photosynthesis of *Enhalus acoroides* and *Thalassia hemprichii* was calculated based on irradiance availability:

$$P_{net} = \frac{P_{max} \; * \; I_z}{K_m \; + \; I_z} + R$$

where:

P_{net} = net production, mg C per g DW photosynthetic material per hour; O_2 units were converted to C units applying a molar rate of 1 according to an overall equation for photosynthetic reaction (Larcher 1975; Jackson & Jackson 1996)

P_{max} = maximum gross production, mg C per g DW photosynthetic material per hour
I_z = irradiance available at seagrass depth, $\mu E.m^{-2}.s^{-1}$
K_m = irradiance level at half-saturation, $\mu E.m^{-2}.s^{-1}$
R = respiration, i.e., P_{net} at zero irradiance

P_{max}, K_m and R for *Enhalus acoroides* and *Thalassia hemprichii* were based on experimental determinations covering the same geographic area (data of N. Agawin & J.S. Uri as published in Vermaat et al. 1995b) and were assumed constant over the entire calculation period (January 1993 - December 1996).

The correction of I_z was done by:

(1) calculating above water irradiance at any time during the day ($I_{above \; water} = I_{noontime} \sin(t)$), assuming a sinusoidal PAR environment (Kraemer & Alberte 1993); t = 0 to daylength; deriving $I_{noontime}$ from the cloudless daily PAR and the global radiation data obtained from the nearest weather station (Chapter 2). Details of daylength calculations are found in Philippart (1995);

(2) correcting for surface reflectance (assumed 10% in this study; see also Scheffer et al. 1993, Hillman et al. 1995) to get $I_{below\ water\ surface}$;

(3) correcting for water depth (datum chart depth + tide level) and turbidity (mean K_d, Chapter 2) using the Lambert-Beer equation. The corresponding tide level at any time of the day was interpolated using a cosine function based on the predicted high and low levels (tide tables, National Mapping and Resource Information Authority, NAMRIA, Philippines).

P_{net} was iterated (at 7-minute steps) with appropriately corrected irradiance (I_z) over 24-hr periods. P_{net} values were then expressed as percentages (mg carbon synthesized per mg carbon biomass of photosynthetic material) in order to have comparable values with the measured relative growth rate (RGR).

Contribution analysis
On a per site basis where surface reflectance, datum depth and turbidity may be assumed approximately constant over time (see also Chapter 2), the total change in the net photosynthesis were decomposed as follows:

$$\Delta P_{net} = \Delta I_s \frac{\partial P_{net}}{\partial I_s} + \Delta C \frac{\partial P_{net}}{\partial C} + \Delta Z_t \frac{\partial P_{net}}{\partial Z_t}$$

(1) (2) (3)

where:

ΔP_{net}	= the total change in P_{net}
ΔI_s	= change in cloudless solar irradiance
ΔC	= change in transmittance through clouds
ΔZ_t	= change in tide level
(1), (2), (3)	= partial contributions of the different terms to ΔP_{net}

These three partial terms (1), (2) and (3) may then be regressed against the total change (ΔP_{net}) to determine the contribution of each factor to the total variations.

Data analyses

AGR and TSA values were log-transformed overcoming both normality and variance heterogeneity problems (Bartlett's and Cochran's C-tests). For RGR, these problems were overcome after inverse arcsine transformation. All statistical tests were done using transformed values, following the procedures described in Sokal and Rolff (1981) and the SPSS/PC+ statistical package (Norusis 1986). Main and interactive effects of sites and sampling months on the variables AGR, TSA and RGR were tested using two-way ANOVA. Tests on the overall means across sites were done using oneway ANOVA with subsequent multiple comparisons tests (Tukey). Linear regression was done to test the strength of the relation between variables, e.g., TSAs vs. grand mean daily PAR values at seagrass depths; AGR vs. TSA; RGR vs. calculated photosynthetic rate; total variations in photosynthetic rate (ΔP_{net}) against its components.

Table 3.1. Two-way ANOVA for the effects of site and sampling months on the total leaf surface area (TSA) and relative growth rate (RGR) of *Enhalus acoroides* and *Thalassia hemprichii*. Asterisks (*) attached to F-values indicate significance at $p < 0.001$. Variance component values indicate the variance percentage explained by the corresponding source of variation.

Source of variation	df	MS	F	Variance Component
1. ***Enhalus acoroides***				
TSA (log-transformed)				
Site	5	51.672	594.284*	44.84%
Month	23	2.290	26.341*	7.66%
Site x Month	98	0.624	7.173*	11.20%
within samples	2202	0.087		36.30%
RGR (arcsin-transformed)				
Site	5	9.014	268.485*	21.24%
Month	23	2.531	75.394*	23.63%
Site x Month	98	0.325	9.668*	16.52%
within samples	2203	0.034		38.61%
2. ***Thalassia hemprichii***				
TSA (log-transformed)				
Site	5	8.217	77.097*	19.25%
Month	10	2.916	27.360*	12.22%
Site x Month	48	0.592	5.557*	12.66%
within samples	1391	0.107		55.87%
RGR (arcsin-transformed)				
Site	5	3.408	38.218*	7.64%
Month	10	8.878	99.548*	37.09%
Site x Month	48	0.492	5.519*	10.20%
within samples	1391	0.089		45.07%

RESULTS

Spatio-temporal variation

Shoot size and leaf growth rates of both *Enhalus acoroides* and *Thalassia hemprichii* significantly varied across sites and across sampling months (Table 3.1). For *Enhalus*, the effect of site on the shoot size (variance component: 45%) was stronger than the temporal effect (variance component: 8%) For *Thalassia*, the temporal and spatial effects on shoot size were approximately similar (12 and 19%, respectively). For the relative growth rate (RGR), stronger temporal effect was present in *Thalassia*: the temporal variance components accounted for 37% (vs. 8% for site component) of the overall RGR variance. For *Enhalus*, the contribution of spatial and temporal effects on RGR was comparable (21 and 24%, respectively).

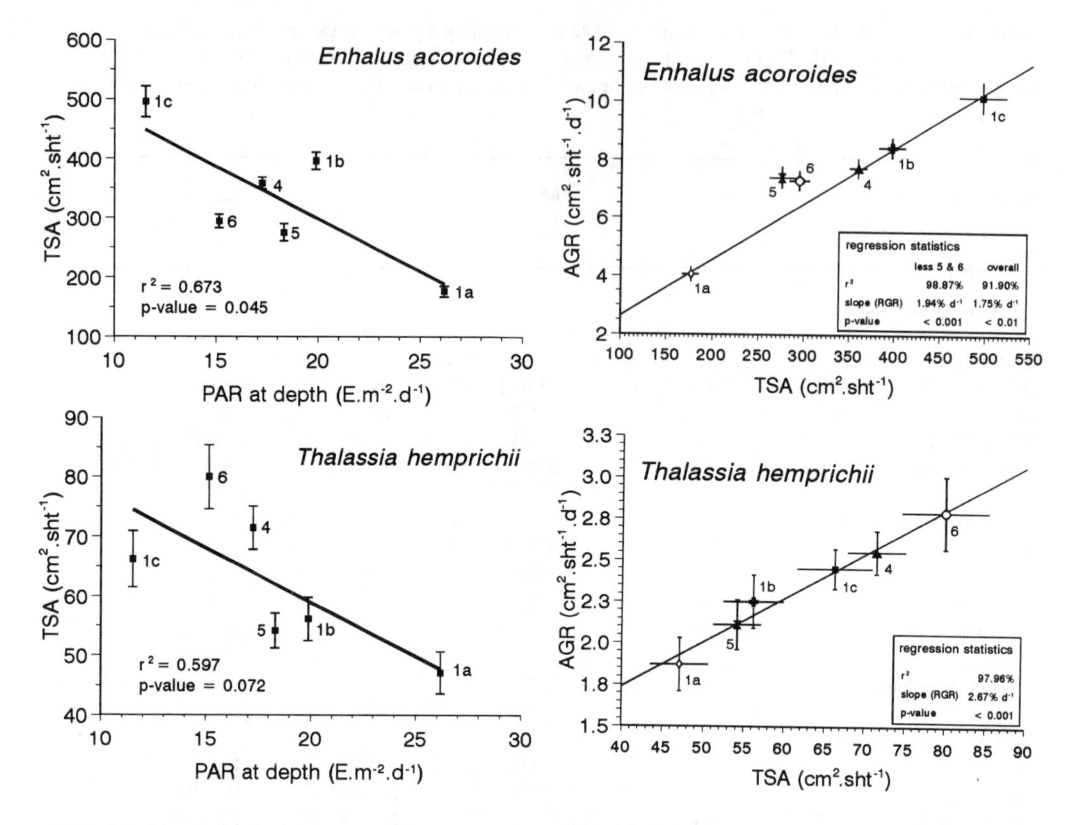

Fig. 3.1. Mean shoot sizes (one-sided total leaf surface area, TSA) of *Enhalus acoroides* and *Thalassia hemprichii* at sites 1a-6 plotted against the mean PAR available at depth. Vertical bars are standard errors.

Fig. 3.2. The absolute leaf growth rates (AGR) of *Enhalus acoroides* and *Thalassia hemprichii* at sites 1a-6 plotted against the corresponding shoot sizes (TSA). Vertical and horizontal lines are standard errors.

The overall average shoot sizes (TSA) of both species were generally larger in darker environments, i.e., in deeper and/or more turbid waters (Table 3.2; Fig. 3.1). In absolute terms, larger plants produced significantly more (AGR, Table 3.2; Fig. 3.2) while their relative rates (RGR) remained comparable across sites (Table 3.2; Fig. 3.2). Contrary to *Thalassia*, *Enhalus* showed significant differences between sites in RGR, being significantly higher in shallow, muddy, near-shore sites 5 and 6 (Table 3.2). Overall, the larger *Enhalus* had significantly lower RGR than *Thalassia* (Table 3.2).

The strong influence of sampling months on RGR for both species (Table 3.1) was mainly due to the high variability in RGR among months (i.e., within but also between years) rather than a clear general annual seasonality (Figs. 3.3, 3.4). RGR values in the same months in different years, or, in adjacent months in the same year were often significantly different. Nevertheless, for *Enhalus acoroides*, generally lower (Tukey test, $p < 0.05$) RGR mean values were shown in the period May-August (the rainy season; Fig. 3.3) when the data points were pooled arbitrarily in three "seasons" of four months in a year. This minimum in RGR in May-

Table 3.2. Overall mean (± se) absolute leaf growth rate (AGR), total leaf surface area (TSA) and relative growth rate (RGR) of *Enhalus acoroides* and *Thalassia hemprichii* as measured at the respective sites 1a - 6. Different letters (superscript) attached to values indicate sitewise significant differences at $p < 0.05$ (Tukey test); ns = not significant. Significant differences between overall species means are indicated in the same way.

SITE	n (months)	AGR ($cm^2.sht^{-1}.d^{-1}$)	TSA ($cm^2.sht^{-1}$)	RGR ($\%.d^{-1}$)
1. ***Enhalus acoroides***				
1a	20	4.06 ± 0.24^{a}	176.91 ± 9.00^{a}	2.31 ± 0.07^{ab}
1b	24	8.44 ± 0.34^{bc}	396.42 ± 14.36^{c}	2.15 ± 0.05^{a}
1c	23	10.19 ± 0.51^{c}	495.62 ± 25.83^{d}	2.09 ± 0.05^{a}
4	19	7.73 ± 0.32^{b}	359.02 ± 9.61^{c}	2.16 ± 0.08^{a}
5	22	7.39 ± 0.35^{b}	275.53 ± 14.43^{b}	2.75 ± 0.09^{c}
6	19	7.29 ± 0.33^{b}	294.31 ± 11.21^{b}	2.57 ± 0.11^{bc}
species overall	6 (sites)	7.52 ± 0.82^{b}	332.97 ± 44.84^{b}	2.34 ± 0.11^{a}
2. ***Thalassia hemprichii***				
1a	11	1.87 ± 0.16^{a}	47.11 ± 3.53^{a}	3.90 ± 0.19^{ns}
1b	13	2.25 ± 0.16^{ab}	56.15 ± 3.69^{a}	3.92 ± 0.14^{ns}
1c	13	2.45 ± 0.12^{ab}	66.21 ± 4.68^{b}	3.82 ± 0.15^{ns}
4	13	2.55 ± 0.13^{b}	71.44 ± 3.61^{b}	3.62 ± 0.13^{ns}
5	13	2.11 ± 0.15^{ab}	54.16 ± 2.95^{a}	3.83 ± 0.18^{ns}
6	14	2.79 ± 0.22^{b}	79.97 ± 5.41^{b}	3.44 ± 0.14^{ns}
species overall	6 (sites)	2.34 ± 0.13^{a}	62.51 ± 4.99^{a}	3.76 ± 0.08^{b}

August period is particularly the case for shallow, turbid and muddy sites 5 and 6 where RGR mean values were generally high but between-month RGR variation was large (Fig. 3.3). For *T. hemprichii*, such differences between pooled "seasonal" means could not be detected (Fig. 3.4), although here also, extremely low RGR values were observed one period in the rainy season (late May 1995).

Comparison between P_{net} and RGR

As calculated here, the photosynthetic rate (P_{net}) represents the total carbon production rate while RGR represents the carbon allocation to leaf growth. In other words, the realized leaf RGR is the remainder of the gross production and subsequent allocation to other plant parts as well as their respiration. Therefore, the ratio (RGR/P_{net}) approximates the relative energy allocation to leaf production. For both *Enhalus acoroides* and *Thalassia hemprichii*, this ratio was greater than 70% indicating a substantial amount of energy allocated to leaf production.

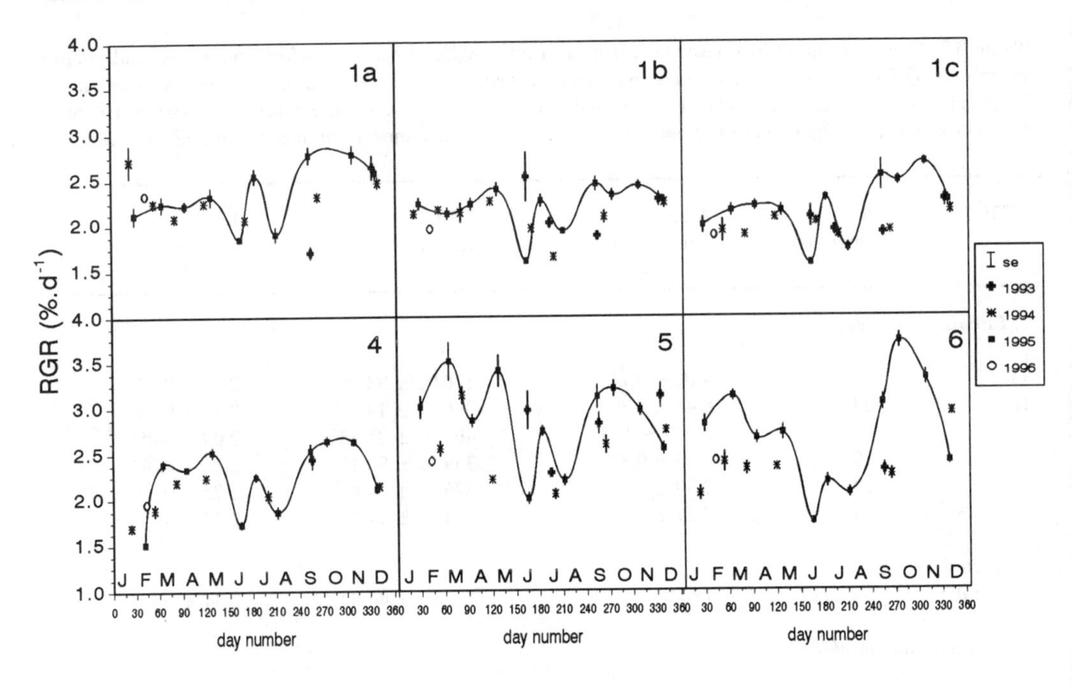

Fig. 3.3. Seasonal variation in the relative leaf growth rates (RGR) of *Enhalus acoroides* at sites 1a-6. Vertical bars are standard errors.

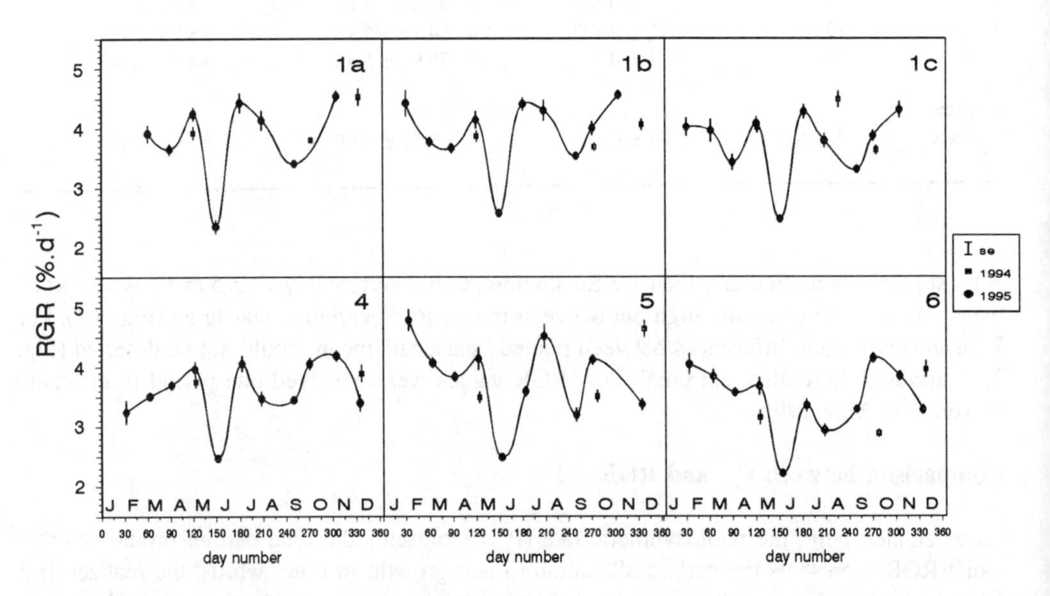

Fig. 3.4. Seasonal variation in the relative leaf growth rates (RGR) of *Thalassia hemprichii* at sites 1a-6. Vertical bars are standard errors.

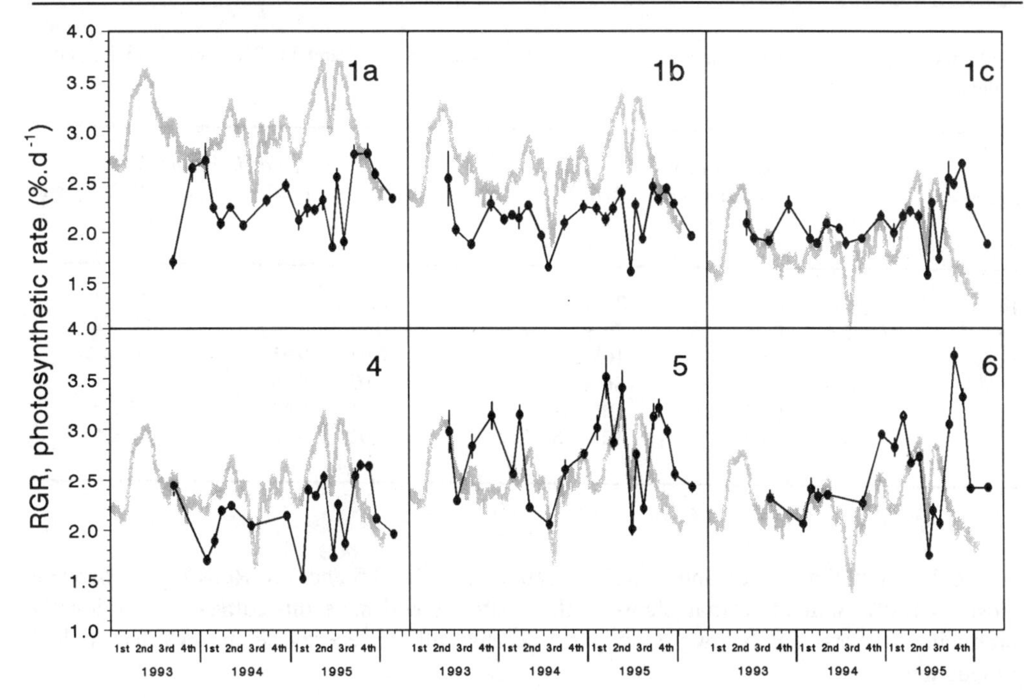

Fig. 3.5. Comparison between the calculated photosynthetic rates (P_{net}, smoothed curves) and the relative growth rates (RGR) of *Enhalus acoroides* at sites 1a-6; linear statistics r^2 and p-values comparing the two curves are shown in Table 3.3.

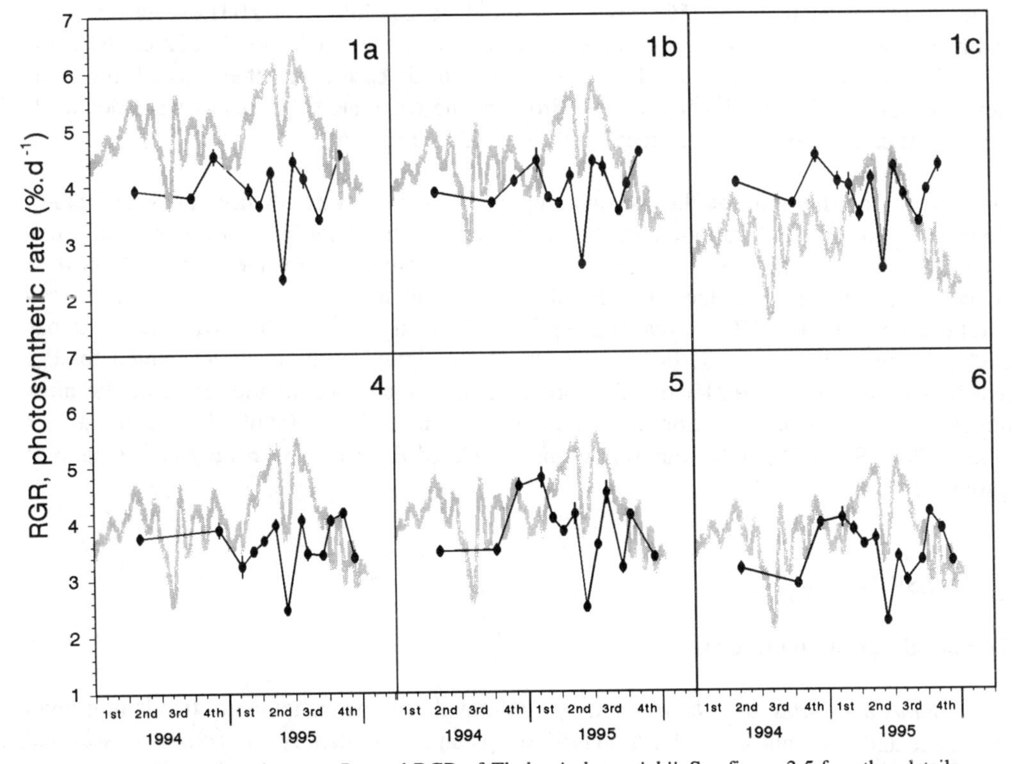

Fig. 3.6. Comparison between P_{net} and RGR of *Thalassia hemprichii*. See figure 3.5 for other details.

Table 3.3. Regression statistics comparing P_{net} and RGR shown in Figs. 3.5 and 3.6. The mean RGR/P_{net} ratios at the sites are also shown.

SITE	*Enhalus acoroides*			*Thalassia hemprichii*		
	R^2 (%)	p-value	RGR/P_{net} ratio (%)	R^2 (%)	p-value	RGR/P_{net} ratio (%)
1a	6.0	0.42	74 ± 3	0.11	0.92	72 ± 4
1b	35.6	0.01	79 ± 2	0.17	0.90	82 ± 4
1c	34.4	0.02	105 ± 5	0.04	0.95	107 ± 7
4	42.6	0.03	86 ± 4	0.09	0.76	82 ± 4
5	25.0	0.07	106 ± 4	1.61	0.68	85 ± 5
6	31.0	0.07	108 ± 6	0.02	0.96	85 ± 5

At the darker stations (deep and/or turbid; Table 3.3; Figs. 3.5 and 3.6) RGR/P_{net} ratios were close to 100%, indicating that plants at these sites would have difficulties in additionally producing flowers and belowground parts because all the energy seems devoted to leaf production.

The calculated photosynthetic rates (P_{net}) for *Enhalus acoroides* based on PAR availability at the sites generally correlated with the measured RGR (Fig. 3.5; Table 3.3). In most sites, R^2 values were greater than ca. 25% (highest at site 4, R^2 ca. 43%, p = 0.03) except the weak and insignificant correlation at a shallow and clear water site (1a, Table 3.3; R^2 ca. 6%; p = 0.42). For *Thalassia hemprichii*, R^2 values were, in all cases, less than 2% (Table 3.3). Nevertheless, in May 1995 when drastic drops in the calculated P_{net} values were obtained, parallel drops in RGR were also measured in all sites (Fig. 3.6).

The partial contributions of the major components (clear-sky irradiance, tides and cloud attenuance) to the total variation in the calculated P_{net} for *Enhalus acoroides* in all sites generally follow similar trends as shown in Fig. 3.7. Relative to the value at an arbitrary date (1 January 1993), the changes in clear-sky solar irradiance contributed to P_{net} increases ranging from 0 to +0.51%.d^{-1} (ca. 0-2 mgC per 1000 mg DW of photosynthetic parts per day) peaking at ca. mid-year. In contrast, the fluctuations in tide levels accounted for P_{net} reductions ranging 0 to -0.21%.d^{-1}. The cloud attenuance component showed to be the most important factor, the contribution of which ranged from -1.5% to +0.66%.d^{-1}, accounting for at least 89% (R^2) of the total variation in the calculated P_{net} for *Enhalus acoroides* (Fig. 3.8; Table 3.4).

DISCUSSION

Temporal variations in growth

For *Enhalus acoroides* and *Thalassia hemprichii*, explanations for temporal trends in growth are scarce and often unclear. Johnstone (1979) was unable to identify any consistent seasonal

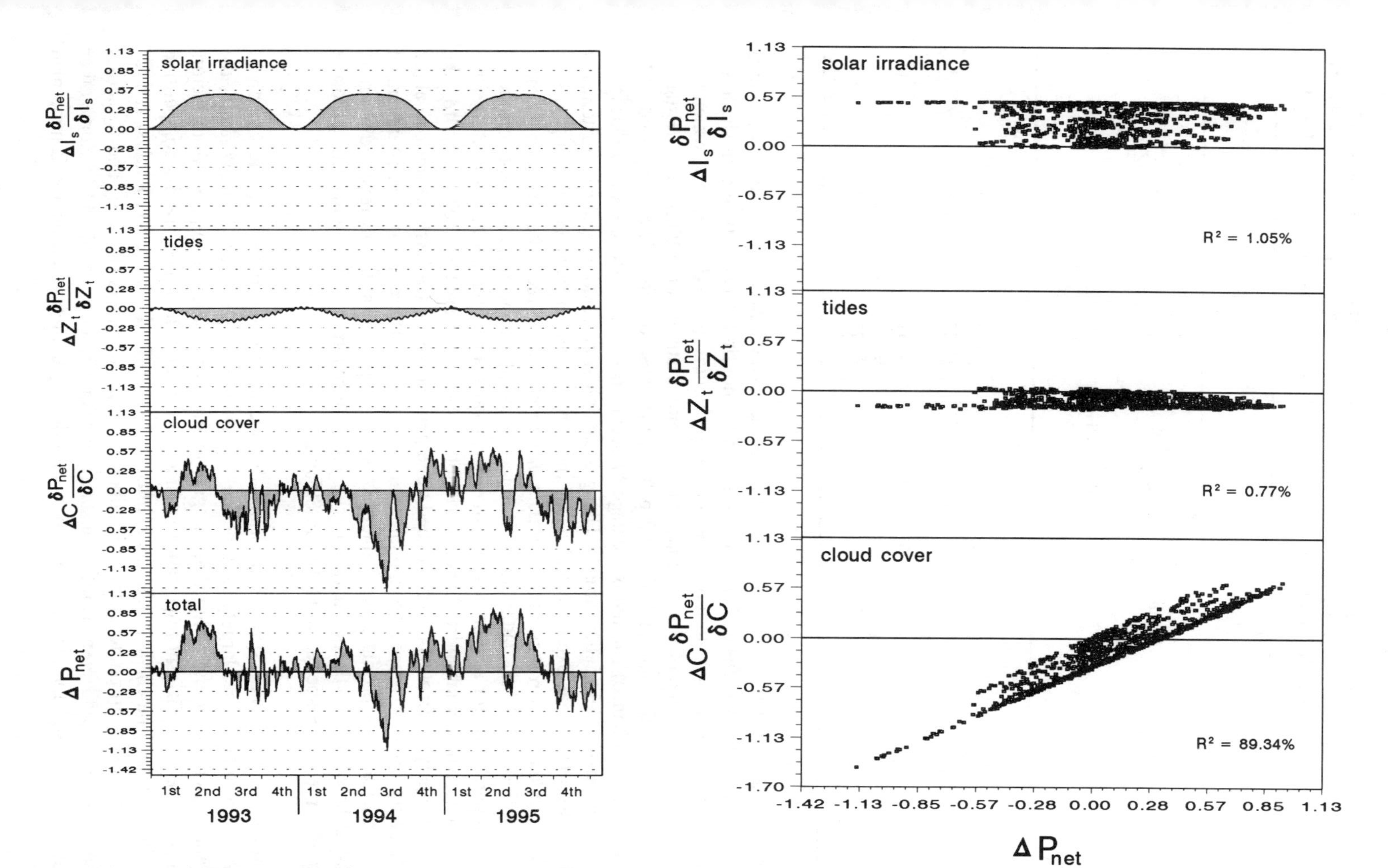

Fig. 3.7. Partial contributions (%.d^{-1}) of the clear-sky solar irradiance, cloud attenuation and tides to the total variations in photosynthesis at site 1a. See Materials and Methods for respective calculations.

Fig. 3.8. Partial contributions (due to solar irradiance, tidal fluctuations and cloud cover) regressed against the total variation (ΔP_{net}).

Table 3.4. R^2 values of the separate linear regressions of ΔP_{net} against its components (see Materials and Methods) for the six study sites. All R^2 values are significant at $p < 0.05$ except those with attached 'ns'. Datum depths (m) and mean extinction coefficients (K_d, m^{-1}) are also shown. See Fig. 3.8 for the processing of the data of site 1a.

			R^2 (%)		
SITE	depth (m)	K_d (m^{-1})	$\Delta I_s \frac{\partial P_{net}}{\partial I_s}$ solar irradiance	$\Delta C \frac{\partial P_{net}}{\partial C}$ clouds	$\Delta Z_t \frac{\partial P_{net}}{\partial Z_t}$ tides
1a	0.35	0.31	1.05	89.34	0.77
1b	1.25	0.31	1.96	89.03	1.48
1c	3.0	0.31	4.42	88.57	3.43
4	1.2	0.41	0.28^{ns}	94.19	0.18
5	0.58	0.60	0.26^{ns}	97.71	0.52
6	0.70	0.72	0.81	98.75	1.22

pattern in his 2-year data set of *Enhalus acoroides* leaf growth rates. In the same study area (Papua New Guinea), Brouns and Heijs (1986) also did not find seasonality in their shallower station but found a seasonal, unimodal trend in deep *Enhalus acoroides* meadow which they attributed to the influence of water temperature. A similar study in the Philippines (Estacion & Fortes 1988) found a seasonal bimodal pattern, the cause of which was suggested to be daylength (i.e., irradiance) and the percentage of occurrence of the low low water during the day (i.e., immersion-dessication stress). Using a canonical correlation analysis, Erftemeijer (1993) showed that tidal exposure and water movement were primary determining growth factors in *Enhalus acoroides* and *Thalassia hemprichii* at a reef site (Barang Lompo, Sulawesi, Indonesia). In the latter study (Erftemeijer 1993), temperature was not considered to be an important factor as there was no significant variation of temperature found over the study period, but available irradiance was not among the variables considered for the multivariate analysis. Brouns (1985) studied three populations of *Thalassia hemprichii* along a depth gradient (0.4, -0.3 and -0.5 m relative to chart datum). There was no consistent annual trend for all sites except that a more intensive study of growth at 0.4 m revealed a strong correlation between growth and the tidal cycle, with reduced growth during low tide exposure as also observed by Stapel et al. (1996) in Sulawesi.

The present study showed strong indications that the temporal variations in the growth of *Enhalus acoroides* were largely a function of available light. For *T. hemprichii*, the correlations were weak indicating that the growth function is more complex. More factors which could be of significant importance include (1) the intra- and inter-species competition (crowding, etc.), (2) patch/shoot age (as *Thalassia* is short-lived relative to *Enhalus*) and (3) significant variability in photosynthetic-irradiance curve parameters (P_{max}, K_m and R) across time and space due to differences in, for instance, nutrients (Agawin et al. 1996). Some other seagrasses, e.g., *Zostera marina* are known to vary these parameters seasonally (Zimmerman et al. 1995) and spatially (Vermaat et al. in press; Dutch vs. Spanish populations). The following conclusions can be drawn for *Enhalus acoroides*:

1.) *Shallow vs. deep populations in clear waters.* Leaf growth of *Enhalus acoroides* in shallow, clear water sites (e.g., 1a) is less sensitive to temporal variation in light than in deep, clear water sites (Table 3.3, Fig. 3.5; Brouns & Heijs 1986; Johnstone 1979), because light levels at the shallower sites probably lie along the asymptotic range of the photosynthesis-irradiance curve (see also Chapter 8).

2.) *Cloud cover impact.* The effect of cloud cover attenuance on temporal variation in estimated net photosynthesis by far exceeded that of other components (e.g., cloudless solar irradiance, tides; Table 3.3). The fact that cloud cover (i.e., the weather) is highly variable basically explains the inconsistent growth trends obtained in various studies. Growth patterns may be generally bimodal (two major drops/peaks, Estacion & Fortes 1988; Vermaat et al. 1995a; this study) when peak cloud formation generally occurs close to 'summer' months. In the Philippines, this usually coincides with the rain and storm peaks (June-September period; Figs. 2.11 and 2.12; see also Kintanar 1986). In contrast, a growth curve is probably unimodal when peak cloud formation occurs towards the 'winter' months, or, when cloud formation is evenly distributed throughout the year. This was probably the case in the studies of Brouns and Heijs (1986) and Erftemeijer (1993), although a seasonal low-tide exposure desiccation during August-December period might strengthen the unimodality (e.g. Erftemeijer 1993).

3.) *Temperature.* The correlation between growth and temperature patterns may be significant (as demonstrated in Brouns & Heijs 1986). The temperature data shown in Chapter 2 (Fig. 3.3) which showed a bimodal annual pattern may also indicate a significant correlation. However, the causality of the relation is doubtful. Firstly, the range of the temperature variation in the tropics is narrow (24-31°C Papua New Guinea, Brouns & Heijs 1986; 26-35°C in Bolinao, Chapter 2; ca. 27°C with only slight seasonal variations, Indonesia, Erftemeijer 1993), thus, may not strongly cause significant growth fluctuations. Secondly, temperature (having relatively smooth curves) could not account for the significant growth variations over short periods. Finally, a strong correlation between temperature and growth should be probable simply because temperature correlates strongly with irradiance (see also Van Tussenbroek 1995).

4.) *Tides.* In view of the previous results (Chapter 2) that the timing and levels of high and low tides are not the same every day, month or year, the net effect is difficult to establish by simple regression. The timing and levels of tides have effects on the total amount of irradiance reaching the plants. Thus when finer (e.g. daily) time-scale growth studies are done, a lunar trend may be found (e.g., Brouns 1985 for *Thalassia hemprichii*), i.e., higher growth in periods with lower water level at daytime and vice versa. And since the level and timing of tides are synodically shifting, also a synodic trend in growth over a lunar cycle may be observed (as probably is the case in Brouns 1985). However, the ultimate cause is probably still be the irradiance availability at depth as affected by timing and levels of tides.

Spatial differences in shoot size and growth

Paradoxically, plants in darker environments formed higher above-ground biomass (Fig. 3.1 and 3.2), and also had higher absolute growth rates (Figs. 3.3 and 3.4; see also Brouns & Heijs 1986 and Estacion & Fortes 1988). This finding however is consistent with the results of the *in situ* and laboratory germination experiments (Chapter 6). Seedlings grown in darker field environments (sites 5 and 6) and those grown in shaded tanks in the laboratory produced bigger plants. The finding in this study that the relative leaf growth rates (RGRs) of *Enhalus acoroides* and *Thalassia hemprichii* remain generally constant across sites has a major implication to the energy allocation of these plants to other functions, e.g., flower and below-ground production. Considering that the amount of energy (photosynthates) decreases as the amount of light decreases, the allocation to other parts (flowers, roots and rhizomes) should decrease to maintain the same allocation effort for leaves. Apparently, in the present seagrasses, leaf production has a priority over e.g. sexual reproduction. Thus, it might be hypothesized that *Enhalus acoroides* and *Thalassia hemprichii* plants growing in the darker environments produce less flowers and/or below-ground biomass. The former is confirmed by the fact that flowering events are rare in deep populations (Chapter 5).

LITERATURE CITED

Agawin, N.S.R., C.M. Duarte and M.D. Fortes. 1996. Nutrient limitation of Philippines seagrasses (Cape Bolinao, NW Philippines): *in situ* experimental evidence. Mar. Ecol. Prog. Ser. 138: 233-243.

Barber, B.J. and J.P. Behrens. 1985. Effects of elevated temperature on seasonal *in situ* leaf productivity of *Thalassia testudinum* Banks ex König and *Syringodium filiforme* Kützing. Aquat. Bot. 22: 61-69.

Brouns, J.J.W.M. 1985. A comparison of the annual production and biomass in three monospecific stands of the seagrass *Thalassia hemprichii* (Ehrenb.) Aschers. Aquat. Bot. 23: 149-175.

Brouns, J.J.W.M. 1987a. Aspects of production and biomass of four seagrass species (Cymodoceoideae) from Papua New Guinea. Aquat. Bot. 27: 333-362.

Brouns, J.J.W.M. 1987b. Growth patterns in some Indo-West Pacific seagrasses. Aquat. Bot. 28: 39-61.

Brouns, J.J.W.M. and H.M.L. Heijs. 1986. Production and biomass of the seagrass *Enhalus acoroides* (L.f.) Royle and its epiphytes. Aquat. Bot. 25: 21-45.

Bulthuis, D.A. 1987. Effects of temperature on photosynthesis and growth of seagrasses. Aquat. Bot. 27: 27-40.

Erftemeijer, P.L.A. 1993. Factors limiting growth and production of tropical seagrasses: nutrient dynamics in Indonesian seagrass beds. PhD thesis, Katholieke Universiteit Nijmegen, The Netherlands, 173 pp.

Estacion, J.S. and M.D. Fortes. 1988. Growth rates and primary production of *Enhalus acoroides* (L.f.) Royle from Lag-it, North Bais Bay, The Philippines. Aquat. Bot. 29: 347-356.

Frenette, J.J., S. Demers, L. Legendre and J. Dodson. 1993. Lack of agreement among models for estimating the photosynthetic parameters. Limnol. Oceanogr. 38: 679-687.

Hillman, K., A.J. McComb and D.I. Walker. 1995. The distribution, biomass and primary production of the seagrass *Halophila ovalis* in the Swan/Canning Estuary, Western Australia. Aquat. Bot. 51: 1-54.

Hootsmans, M.J.M. and J.E. Vermaat. 1994. Light response curves of *Potamogeton pectinatus* L. as a function of plant age and irradiance level during growth. In: W. van Vierssen, M.J.M. Hootsmans and J.E. Vermaat (eds.). Lake Veluwe, a macrophyte-dominated system under eutrophication stress. Geobotany, Kluwer Academic Publishers, Dordrecht, The Netherlands, 62-117 pp.

Jackson, A.R.W. and J.M. Jackson. 1996. Environmental science. The natural environment and human impact. Longman Group Ltd., Singapore, 370 pp.

Jassby, A.D. and T. Platt. 1976. Mathematical formulation of the relationship between photosynthesis and light for phytoplankton. Limnol. Oceanogr. 21: 540-547.

Johnstone, I. 1979. Papua New Guinea seagrasses and aspects of the biology and growth of *Enhalus acoroides* (L.f.) Royle. Aquat. Bot. 7: 197-208.

Kintanar, 1986. Climatological normals. PAG-ASA, The Philippines.

Kraemer, G.P. and R.S. Alberte. 1993. Age-related patterns metabolism and biomass in subterranean tissues of *Zostera marina* (eelgrass). Mar. Ecol. Prog. Ser. 95: 193-203.

Larcher, W. 1975. Physiological plant ecology (translated edition). Springer-Verlag, Berlin, 252 pp.

Meñez, E.G., R.C. Phillips and H. Calumpong. 1983. Seagrasses from the Philippines. Smithsonian Contrib. Mar. Sci. 21, 40 pp.

Norusis, M.J. 1986. SPSS-PC+ manual. SPSS Inc. Chicago.

Philippart, C.J.M. 1995. Seasonal variation in growth and biomass of an intertidal *Zostera noltii* stand in the Dutch Wadden Sea. Neth. J. Sea. Res. 33: 205-218.

Santamaria-Galdon, L.E. and W. van Vierssen. 1995. Interactive effect of photoperiod and irradiance cycle of a winter annual waterplant *Ruppia drepanensis* Tineo. In: L.E. Santamaria-Galdon. The ecology of *Ruppia drepanensis* Tineo in Mediterranean brackish marsh (Doñana Park, SW Spain) - a basis for the management of semiarid floodplain wetlands. PhD dissertation, IHE-LUW, The Netherlands, 15-87 pp.

Scheffer, M., A.H. Bakema and F.G. Wortelboer. 1993. MEGAPLANT: a simulation model of the dynamics of submerged plants. Aquat. Bot. 45: 341-356.

Sokal, R. and F.J Rohlf. 1981. Biometry. The principles and practice of statistics in biological research, 2nd ed. WH Freeman and Co., New York, 869 pp.

Stapel, J., R. Manuntun and M. Hemminga. 1996. Biomass loss and nutrient redistribution in an Indonesian *T. hemprichii* seagrass bed following seasonal low tide exposure to daylight. In: J. Stapel. Nutrient dynamics in Indonesian seagrass beds: factors determining conservation and loss of nitrogen and phosphorus. PhD Dissertation, KUN, The Netherlands, 19-31 pp.

Tomasko, D.A., C.J. Dawes, M.D. Fortes, D.B. Largo and M.N.R. Alava. 1993. Observations on a multi-species seagrass meadow off-shore of Negros Occidental, Republic of the Philippines. Bot. Mar. 36: 303-311.

Van Lent, F., J.M. Verschuure, M.L.J. van Veghel. 1995. Comparative study on populations of *Zostera marina* L. (eelgrass): *in situ* nitrogen enrichment and light manipulation. J. Exp. Mar. Biol. Ecol. 185: 55-76.

Van Tussenbroek, B.I. 1995. *Thalassia testudinum* leaf dynamics in a Mexican Caribbean coral reef lagoon. Mar. Biol. 122: 33-40.

Vermaat, J.E., N.S. Agawin, C.M. Duarte, S. Enriquez, M.D. Fortes, N. Marba, J.S. Uri and W. van Vierssen. In press. The capacity of seagrasses to survive eutrophication and siltation, across-regional comparisons. Ambio.

Vermaat, J.E., N.S.R. Agawin, C.M. Duarte, M.D. Fortes, N. Marba and J.S. Uri. 1995a. Meadow maintenance, growth and productivity of a mixed Philippine seagrass bed. Mar. Ecol. Prog. Ser. 124: 215-225.

Vermaat, J.E., C.M Duarte and M.D. Fortes. 1995b. Latitudinal variation in life history patterns and survival mechanisms in selected seagrass species, as a basis for EIA in coastal marine ecosystems (Final Report). Project EC DG XII-G CI1*-CT91-0952, IHE Delft, The Netherlands, 38 pp. + 2 annexes.

Zieman, J.C. 1974. Methods for the study of growth and production of turtlegrass *Thalassia testudinum*. Aquaculture 4: 139-143.

Zimmerman, R.C., J.L. Reguzzoni and R.S. Alberte. 1995. Eelgrass (*Zostera marina* L.) transplants in San Francisco Bay: Role of light availability on metabolism, growth and survival. Aquat. Bot. 51: 67-86.

Chapter 4

Recolonization in a multi-species seagrass meadow: the contrasting strategies of the two dominant species *Enhalus acoroides* (L.f.) Royle and *Thalassia hemprichii* (Ehrenb.) Aschers.

Abstract. This study shows that in a multi-species seagrass meadow in a shallow and clear water site, all the former seagrass species were able to recolonize in the artificially created gaps of 0.25 m^2 in size within ca. 2 years. Extrapolation of the recolonization curves of the different species predicted a full recovery within 10 years post disturbance. Fitted curves for the dominant species *Enhalus acoroides* and *Thalassia hemprichii* showed contrasting strategies, the latter having a comparatively high intrinsic rate, achieving full recovery within ca. 2 years post-disturbance. *Enhalus acoroides* was the latest species to establish and the projected full-recovery time was among the longest (ca. 10 years). The effect of timing of gap creation was generally not significant (except for *Syringodium isoetifolium*) as the temporal variation in density of most species was not significant either outside the gaps. As recolonization by sexual propagules was found to be of less importance, increasing the gap size would most probably require a much longer recovery period. A crude estimate for *Enhalus acoroides* would be > 10 years for 1 m^2 of gap. Further, since the densities of most seagrass species significantly vary between sites and colonization rates are depending on the neighboring densities, the recovery curves would also be different across sites.

INTRODUCTION

Physical loss of seagrasses has been documented to occur due to a number of disturbance types. Extensive losses due to cyclones and storms are frequently reported (Thomas et al. 1961; Cambridge 1975; Patriquin 1975; Birch & Birch 1984; Williams 1988; Poiner et al. 1989; Tilmant et al. 1994; Preen et al. 1995; Short & Wyllie-Echeverria 1996). In the Philippines, seagrass losses due to typhoons might be considerable considering that, on average, 20 tropical storms hit the country annually (Fortes 1988; Rivera 1997), 3-5 of those traverse across Bolinao (16°N; 120°E, Fig. 2.1; pers. obs. 1993-1995), the study location. Less extensive and less documented disturbances include periodic dredging (Holligan & de Boois 1993; Preen et al. 1995; pers. obs., marina construction), dugong-cleared patches (De Iongh et al. 1995; Preen 1995; pers. obs., Northern Palawan, Philippines) and bioturbation (Suchanek 1983; Valentine et al. 1994).

Moderate disturbance may maintain the diversity in space-limited communities (Paine & Levin 1981; Sousa 1984; Barrat-Segretain & Amoros 1995; Holt et al. 1995; Chiarello & Barrat-Segretain 1997). However, the understanding of the population dynamics within the habitats subject to disturbance requires detailed description of (1) disturbance regime, distribution and timing and (2) the patterns of colonization and succession within patches (Sousa 1984). Plant communities such as seagrasses may recolonize cleared patches, or gaps, both by dispersal of sexual propagules and/or vegetative expansion of plants at the periphery (border effect, Chiarello & Barrat-Segretain 1997).

This chapter presents the patterns of recolonization and succession of seagrasses following artificial disturbance (creating bare patches) in a mixed seagrass meadow. Special focus was given to the recolonization rates of the two dominant species (*Enhalus acoroides* and *Thalassia hemprichii*), and the following questions were addressed: (1) Are these two species able to recolonize cleared areas within 2 years?; and (2) How much time does it take to reach the original densities after a disturbance in relation to patch size?.

In connection with the artificial disturbance, undisturbed control plots were also studied to obtain information on shoot densities of *Enhalus acoroides* and *Thalassia hemprichii* at different sites. This was done to (1) obtain reference values for the dominant species and (2) to quantify how these reference values vary between sites and seasons.

Using an age reconstruction technique (Duarte et al. 1994), the time-scale of the clonal replication of the long-lived *Enhalus acoroides* was determined. Fitting the above pieces of information together for *Enhalus*, a crude extrapolation has been made on the relation between the size of disturbance, the "peripheral" density (i.e., the abundance along the gap borders) and the time-scale for recovery towards the pre-clearance density.

MATERIALS AND METHODS

Cleared patches

In a mixed seagrass meadow at a shallow and clear-water site (1a; see Fig. 2.1, Chapter 2), gaps were created. In July 1993, three quadrats (50 cm x 50 cm) were randomly chosen and fixed with permanent markers. After counting all seagrass shoots, all plants including roots and rhizomes within the quadrats were removed. Removal was done as carefully as possible, minimizing sediment loss due to resuspension. In November 1993, another three quadrats were cleared as above. This was done to determine whether the seasonal timing of the disturbance significantly affects the patterns of recolonization and species establishment. During each sampling (once per 1-2 months), all new seagrass colonizers were counted. When distinguishable, seedling colonizers (sexual propagules) were separately counted. The intrinsic rates of colonization, r, were then approximated (nonlinear regression, SPSS/PC+; Norusis 1986) for each species by fitting the corresponding data sets into the simple logistic equation (Richards 1969; Verhulst logistic, Hutchinson 1978).

$$N_t = \frac{K}{1 + e^{-rt}}$$

where: N_t = the number of shoots at any time t (days)
K = maximum shoot number (pre-clearance value)
r = intrinsic rate of increase (% per unit time t)

Spatio-temporal variation in shoot densities of *Enhalus acoroides* and *Thalassia hemprichii* in undisturbed plots

At sites 1a, 1c, 4, 5 and 6, three quadrats (50 cm x 50 cm) were randomly chosen. These were fixed with permanent markers for subsequent sampling of the same plots. During each sampling (once per 1-2 months), shoot densities of *Thalassia hemprichii* and *Enhalus acoroides* were counted underwater by superimposing a metal quadrat (50 cm x 50 cm with grids every 10 cm) taking the same orientation every sampling. When distinguishable, seedlings were counted separately. The occurrence of flowers was also noted. Data shown here were collected from January 1994 to September 1995. For site 5, obtaining a complete data set was not possible due to the intensive fishing activities using gillnets and fish corrals. Quadrat markers were often lost and probably removed on purpose, as those might hamper fishing operations.

Intrinsic rate of shoot increase and the rate of elongation of rhizomes in an *Enhalus acoroides* clone: ageing technique

The rate of vegetative shoot production of a clone can be calculated for species like *Enhalus acoroides* because it maintains good rhizome connections. The age reconstruction technique (Duarte et al. 1994) may then be used to determine rhizome elongation rates and shoot propagation (see also Fig. 1.3, Chapter 1). For this purpose, three clones of *Enhalus acoroides* were collected randomly from each of the sites 1a, 1c, 5 and 6. Each of these twelve clones was aged by counting all the rhizome internodes and subsequently multiplying these values with the corresponding plastochrone interval values (PI: 31.2 ± 1.7, 26.9 ± 0.4, 27.1 ± 1.4, 26.2 ± 1.6 respectively for 1a, 1c, 5 and 6; Rollon et al., Unpublished Data). The mean shoot elongation rate was calculated (i.e., rhizome length ÷ rhizome age). The net vegetative growth rates of each clone (i.e., the intrinsic rates of increase in the number of living shoots per unit time; see also Fig. 1.3, Chapter 1) were approximated by regressing nonlinearly the number of living shoots in a clone with clonal age using the same simple logistic equation as above. A maximum K-value of 8 shoots per clone (based on earlier pers. obs.) was assumed for all sites.

Data analyses

Multiple comparisons of different fitted curves were done by calculating an F statistic (see equation below; Vermaat & Hootsmans 1994), subsequently evaluating its significance at $p < 0.05$. The comparisonwise error rate (CER) was adjusted to maintain an experimentwise error rate (EER) of 0.05 according to the relation, $CER = 1 - (1\text{-}EER)^{1/n}$, where n is the total number of comparisons made.

$$F = \frac{(RSS_{1+2} - (RSS_1 + RSS_2)) \;/\; (df_{1+2} - (df_1 + df_2))}{(RSS_1 + RSS_2) \;/\; (df_1 + df_2)}$$

where: RSS_1 = residual sum of squares for dataset 1
RSS_2 = residual sum of squares for dataset 2

RSS_{1+2} = residual sum of squares for the combined datasets
df_1, df_2, df_{1+2} = the corresponding degrees of freedom

The temporal variations (per site basis) in the log-transformed densities of *Enhalus acoroides* and *Thalassia hemprichii* in the permanent quadrats were tested using ANOVA (repeated measures design).

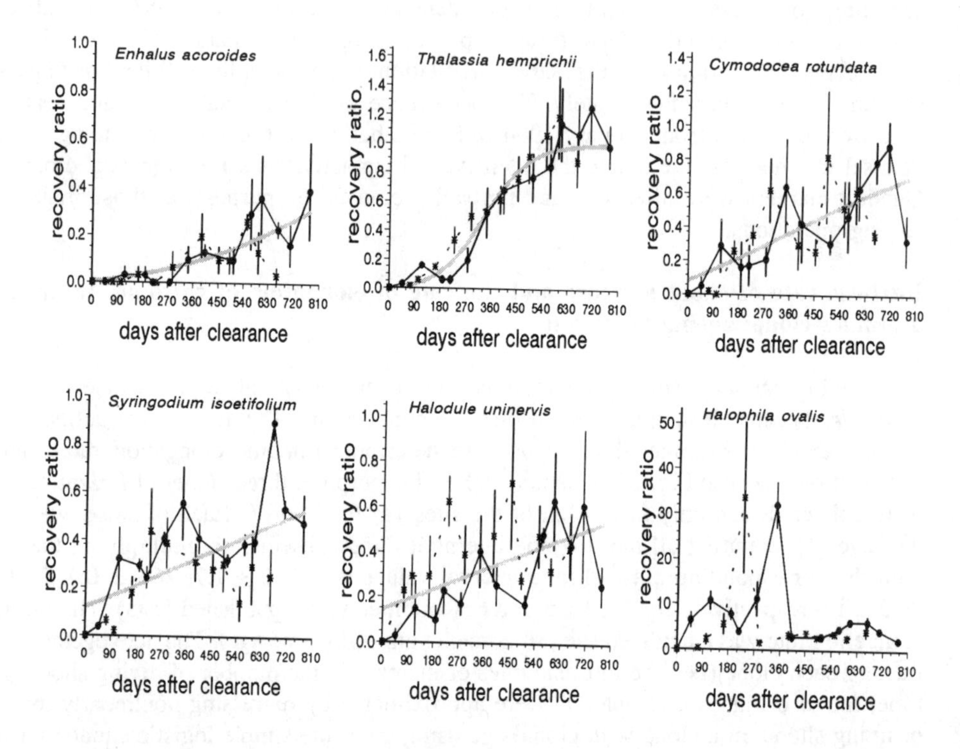

Fig. 4.1. Recolonization by different seagrass species in cleared gaps within a mixed seagrass meadow at a shallow site (1a). Y-axis values are expressed as ratios with respect to the pre-clearance densities (e.g., 1 = 100% recovery: note the the differences in the vertical scale). Solid, thin curves = patches cleared in July 1993; Dotted curves = November 1993 clearing; Error bars = standard errors; thick and smooth curves are nonlinear fits (Verhulst logistic) to the combined datasets (see Table 4.1 for details). The logistics fit for *Halophila ovalis* was poor and is not shown.

RESULTS

Recolonization patterns and succession

All the six seagrass species that were originally present recolonized the gaps, in varying extents, within 2 years (Fig. 4.1). Different species showed different recolonization patterns. The fugitive (i.e., competitively inferior but efficient colonizer; *sensu* Sousa 1984) *Halophila ovalis* was the most conspicuous early colonizer reaching density levels of up to 30-fold higher than the pre-disturbance values within half a year after disturbance (Fig. 4.1). The

Table 4.1. Recolonization rates (r) and the time to reach 50% and 90% recovery relative to pre-clearance shoot densities ($T_{50\%K}$ and $T_{90\%K}$ respectively); R^2 = r-square of the nonlinear regression; July = patches cleared in July 1993; November = patches cleared in November 1993; both = pooled July and November data sets; x = logistic equation did not fit the data well; * = visual approximation; different letters (superscript) attached to data sets indicate significant difference comparing the curve obtained from July clearing with the curve obtained from November clearing; ns = not significant. For specieswise comparisons, see Fig. 4.2.

		regression statistics			
SPECIES	data set	R^2 (%)	r (%.d^{-1})	$T_{50\%K}$ (years)	projected $T_{90\%K}$ (years)
Enhalus acoroides	July^{ns}	75.8	0.331	2.55	4.43
	November^{ns}	37.1	0.132	4.19	9.04
	both	61.8	0.292	3.01	4.93
Thalassia hemprichii	July^{ns}	94.3	1.428	0.97	1.39
	November^{ns}	95.0	1.111	0.82	1.35
	both	93.6	1.260	0.91	1.38
Cymodocea rotundata	July^{ns}	59.2	0.270	0.97	3.50
	November^{ns}	49.5	0.262	0.93	3.45
	both	55.2	0.266	0.95	3.48
Syringodium isoetifolium	July^{a}	55.0	0.191	0.78	4.43
	November^{b}	24.2	0.093	1.82	9.29
	both	41.1	0.161	1.11	5.42
Halodule uninervis	July^{ns}	59.3	0.201	1.48	4.85
	November^{ns}	24.5	0.118	0.15	6.30
	both	36.8	0.148	0.90	5.67
Halophila ovalis	x	x	x	< 0.5*	< 0.5*

coming of the larger species in the later stage reduced *Halophila* density close to its 'undisturbed' level of ca. 60 shoots.m^{-2}. *Cymodocea rotundata*, *Syringodium isoetifolium* and *Halodule uninervis* also showed high recovery values during the early stage, attaining 50% during the first three quarters of the first year (post-disturbance). During this first year, the naturally-dominant species *Enhalus acoroides* and *Thalassia hemprichii* were the latest colonizers (Fig. 4.1; 4.2). However, these dominant species appeared to have steady recolonization curves (with higher "r" values, Table 4.1), the latter achieving 100% recovery after only 2 years (Fig. 4.1). In general, four curve groups (recolonization patterns) can be distinguished (Figs. 4.1, 4.2): (1) *Halophila ovalis*, a fugitive strategist; (2) *Cymodocea rotundata*, *Syringodium isoetifolium* and *Halodule uninervis*, all showing high early densities but with generally low intrinsic rates of density increase ("r" < 0.27%.d^{-1}); (3) *Thalassia hemprichii*, a slow-starter but with a remarkably high "r" (Table 4.1; Fig. 4.2), and; (4) *Enhalus acoroides*, the slowest recolonizer but with slightly higher "r" than the second group.

For the gap size of 0.25 m^2, the projected time it takes to achieve 90% recovery (Fig. 4.2; Table 4.1) varied between species but all values were less than 10 years. The effect of the disturbance timing on the recolonization curves was, in general, not significant (Table 4.1, comparing datasets (1) and (2)) except for *Syringodium isoetifolium*. For *Syringodium*, the curve for the July 1993 clearance had higher "r" (0.191% vs. 0.093%) and shorter time to reach 50% recovery ($T_{50\%K}$ = 285 vs. 666 days).

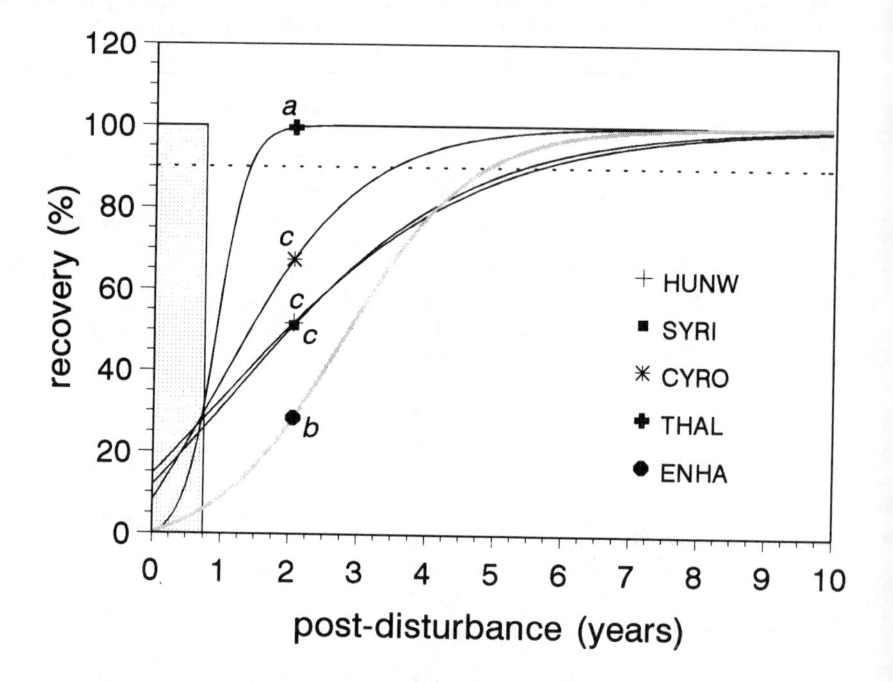

Fig. 4.2. Approximated patterns of recolonization and succession by the five seagrass species: *Enhalus acoroides*; *Thalassia hemprichii*; *Cymodocea rotundata*; *Syringodium isoetifolium*; and *Halodule uninervis*. The pattern for *Halophila ovalis* is not shown. Y-axis depicts ratios with respect to the pre-clearance levels. Horizontal dotted line indicates 90% recovery; shaded bar indicates an initial recolonization period (effectively 3/4 year); different letters attached to curves indicate specieswise significant differences in recolonization curves.

Seedling colonizers

Few sexual propagules (i.e., seedlings) of *Cymodocea rotundata*, *Syringodium isoetifolium* and *Halodule uninervis* were observed. This occurred especially during the early period (July-November 1993), after which distinguishing seedlings from the vegetative recruits of these species was impossible. For the structurally most important species *Enhalus acoroides* and *Thalassia hemprichii*, recolonization by seedlings hardly occurred. Only one seedling of each of the latter two species was able to settle in all six cleared patches over the entire study period. Seedlings of these species can be easily distinguished even when the space is already crowded because they have conspicuously large seeds.

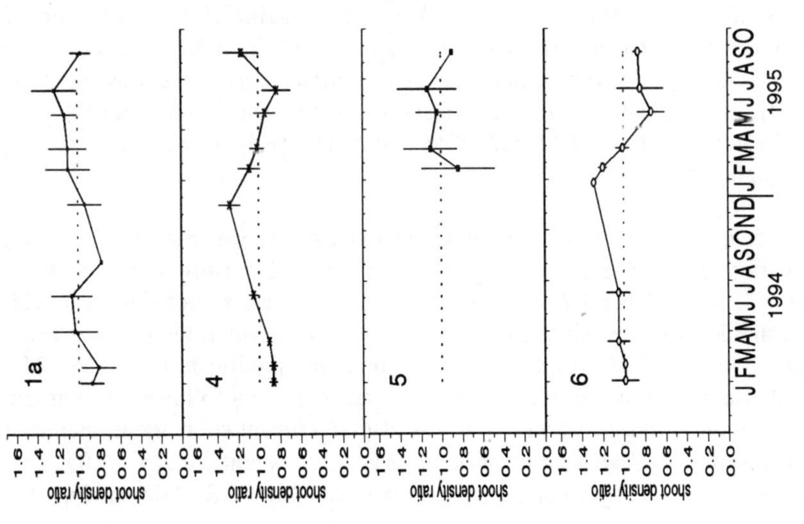

← **Fig. 4.3**. Temporal variation in the density of *Enhalus acoroides* in three undisturbed quadrats (50 cm x 50 cm) in sites 1a, 4, 5 and 6. Y-axis values are expressed as ratios against the overall means (dotted line). There was no *Enhalus acoroides* present in the permanent quadrats in site 1c. For site 5, data before 1995 were not shown because of reasons explained in the Materials and Methods (see also Fig. 4.6).

→ **Fig. 4.4**. Temporal variation in the density of *Thalassia hemprichii* in three undisturbed quadrats (50 cm x 50 cm) in sites 1a, 1c, 4, 5 and 6. For other details, see Fig. 4.3.

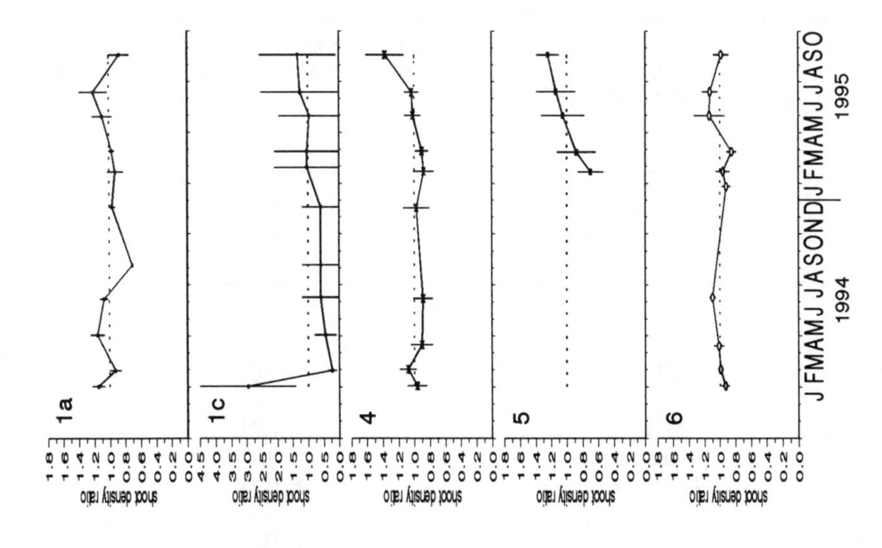

Table 4.2. Shoot densities of *Enhalus acoroides* and *Thalassia hemprichii* in the permanent quadrats. Different letters (superscript) attached to density values indicate sitewise significant differences (log-transformed values) using oneway ANOVA test; x = none present in the permanent quadrats. Testing of temporal variation was done on a per site basis using repeated ANOVA design.

SITE	mean shoot density (shoots per 0.25 m^2)	temporal variation (p-value)
1. *Enhalus acoroides*		
1a	10.48 ± 0.51^{b}	> 0.05
1c	x	x
4	8.87 ± 0.31^{ab}	< 0.05
5	8.83 ± 0.91^{a}	> 0.05
6	15.87 ± 0.49^{c}	> 0.05
2. *Thalassia hemprichi*		
1a	98.48 ± 3.19^{d}	> 0.05
1c	4.52 ± 1.20^{a}	< 0.05
4	43.27 ± 1.90^{c}	> 0.05
5	37.14 ± 8.0^{b}	< 0.05
6	60.30 ± 1.70^{cd}	> 0.05

Spatio-temporal variations in shoot densities of *Enhalus acoroides* and *Thalassia hemprichii* in undisturbed plots

The temporal variation in the shoot density of *Enhalus acoroides* at the shallow, clear-water site (1a) was not significant ($p < 0.05$; Fig. 4.3; Table 4.2) averaging 10.48 ± 0.51 per 0.25 m^2. There was also no strong seasonality found in sites 5 and 6 (Fig. 4.3; Table 4.2). It was only in site 4 where the temporal variation in the density of *Enhalus acoroides* was significant ($p < 0.05$; Table 4.2). Between sites, differences in shoot density were significant (Table 4.2) which appeared to be correlated with PAR reduction due to the overlying water-column (Fig. 4.5) excluding the highest value at site 6.

For *Thalassia hemprichii*, the temporal variation in density at site 1a was not significant (Fig. 4.4; Table 4.2) averaging 98.48 ± 3.19 per 0.25 m^2. In two other sites (4 and 6) where densities were likewise high (Table 4.2), temporal differences were also not significant (Fig. 4.4; Table 4.2). At the deepest site (site 1c), where *Thalassia hemprichii* density was lowest (mean, 4.52 ± 1.20 shoots per 0.25 m^2), temporal variation was significant ($p < 0.05$). Most notable was the collapse during the period January 1994 - March 1994 followed by a gradual recovery (Fig. 4.4). At site 5, temporal variation was also significant showing an increasing trend. Sitewise, the variation in *Thalassia hemprichii* density was significant ($p < 0.05$, Table 4.2) and, similar to *Enhalus*, strongly correlated with the available PAR at depth (Fig. 4.5).

Major events which might have influenced recruitment and mortality of *Enhalus acoroides* and *Thalassia hemprichii* are visualized in Fig. 4.6. Seedlings occurred conspicuously only in sites 1a and 1c. In site 1a, one *Enhalus* seedling inside the permanent quadrat successfully survived (Fig. 4.6, January 1994) while in 1c, *Enhalus* and *Thalassia* seedlings which settled in May 1994 (Fig. 4.6) all died-off later. For sites 4, 5 and 6, no seedlings inside the permanent quadrats were observed during the study period. However for site 6, seed rain probabilities may be assumed high (see mass seed release in July 1994, Fig. 4.6) although seedling establishment at this site may be low (Chapter 6). Except in site 1c, male and female flowers of *Enhalus acoroides* were observed in all sites (for a more detailed discussion on the sexual reproduction of *Enhalus acoroides*, see Chapter 5). Flowers and fruits of *Thalassia hemprichii* were observed only in site 1a. Physical disturbance was observed in site 1c (bioturbation and consequently sediment burial) and in site 5 (intense fishing activity using fish corrals).

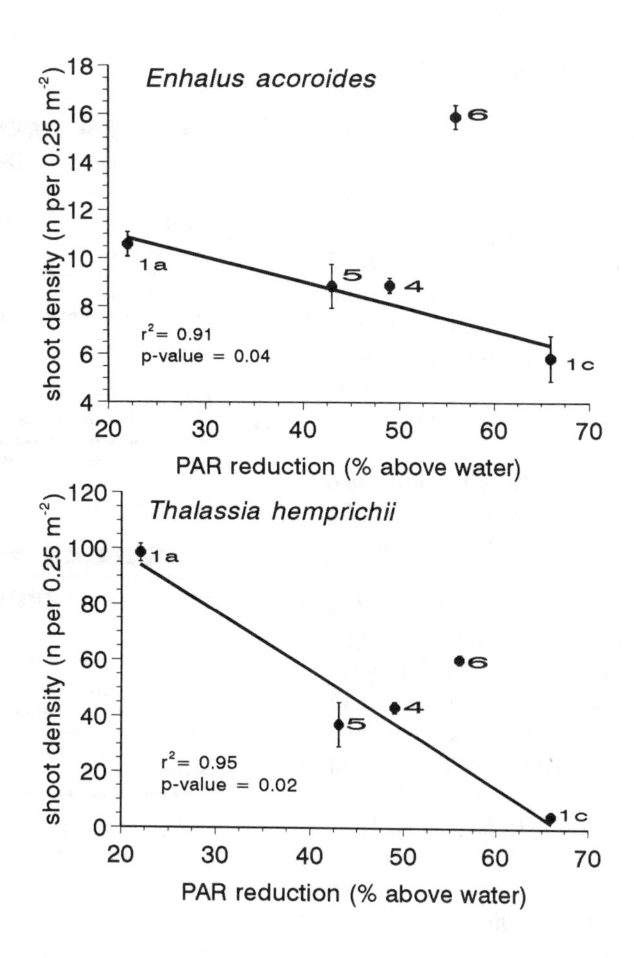

Fig. 4.5. Mean shoot densities of *Enhalus acoroides* and *Thalassia hemprichii* in undisturbed quadrats plotted against mean PAR attenuance due to the overlying water-column. For both species, the regression line excludes site 6 which would otherwise make the line insignificant; note: density of *Enhalus* at site 1c was obtained by counting shoots within 25 random quadrats.

Intrinsic rate of clonal shoot increase and the elongation rate of rhizomes in an *Enhalus acoroides* clone: ageing technique

Relative to other sites, the mean intrinsic rate of shoot increase (r) in the number living shoots within *Enhalus acoroides* clones retrieved from site 1a was low (only ca. 33%.yr^{-1}, Table 4.3; Fig. 4.7). Much higher r values were found in sites 1c, 5 and 6 having 75%.yr^{-1}, 110%.yr^{-1} and 340%.yr^{-1} respectively (Table 4.3; Fig. 4.7). However, the variation between clones collected from the same site was high. For instance, the first *Enhalus* clone at site 1a had two shoots already before the clone was 1-yr old but never again produced another shoot for the next 11 years. The third clone produced the second shoot only after ca. 4 yrs. but after which, the absolute rate of shoot increase was high (ca. 2 $shoots.yr^{-1}$). This high variability between clones is common in most sites, with site 5 being the least variable (Table 4.3; Fig. 4.7).

The rhizome elongation rate of an *Enhalus acoroides* shoot was fastest in site 1c (clonal mean of 7.4 ± 0.62 $cm.sht^{-1}.yr^{-1}$, Table 4.3). The mean annual rhizome elongation in sites 5 and 6 were about the same (4.39 ± 0.15 and 4.89 ± 1.00 $cm.sht^{-1}.yr^{-1}$, respectively) while that of site 1a was relatively low (3.39 ± 0.75 $cm.sht^{-1}.yr^{-1}$).

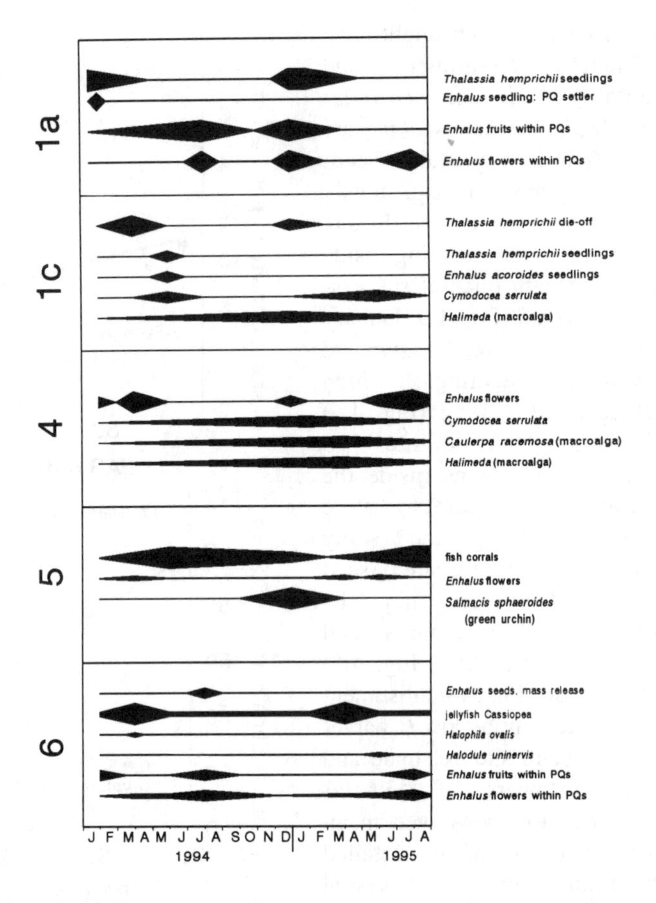

Fig. 4.6. Phenology of the major events observed within the undisturbed plots (PQ) and immediate surroundings. Y-axis scale is arbitrary.

DISCUSSION

In 'undisturbed' multi-species seagrass meadows in shallow waters, the net shoot recruitment of each component species is close to zero, i.e., in approximately equilibrium state. This has been shown to be true for both short-term (e.g., *Enhalus acoroides* and *Thalassia hemprichii* for ca. 2 yrs. in this study) and long-term (age reconstruction technique for all component species in Vermaat et al. 1995) scales. A number of possible mechanisms underlying this semi-steady state can be hypothesized: (1) all species have approximately zero recruitment and zero mortality; (2) recruitment is high but mortality is equally high, and: (3) mortality may be low but further recruitment is space-limited. The implication of the first two possibilities is that when disturbance occurs, e.g., typhoon 'blowouts', recovery would be difficult while the third mechanism implies a rather fast recovery so long as the post-disturbance physical condition is not too harsh (Sousa 1984). In many natural communities, space is a limiting resource (e.g. Paine & Levin 1981; Sousa 1984; Barrat-Segretain & Amoros 1995; Holt et al. 1995; Chiarello & Barrat-Segretain 1997) but evidence for seagrass communities is scarce. This study suggests that, at least, the naturally non-dominant species are space-limited. This is most pronounced for *Halophila ovalis*, taking advantage of the open

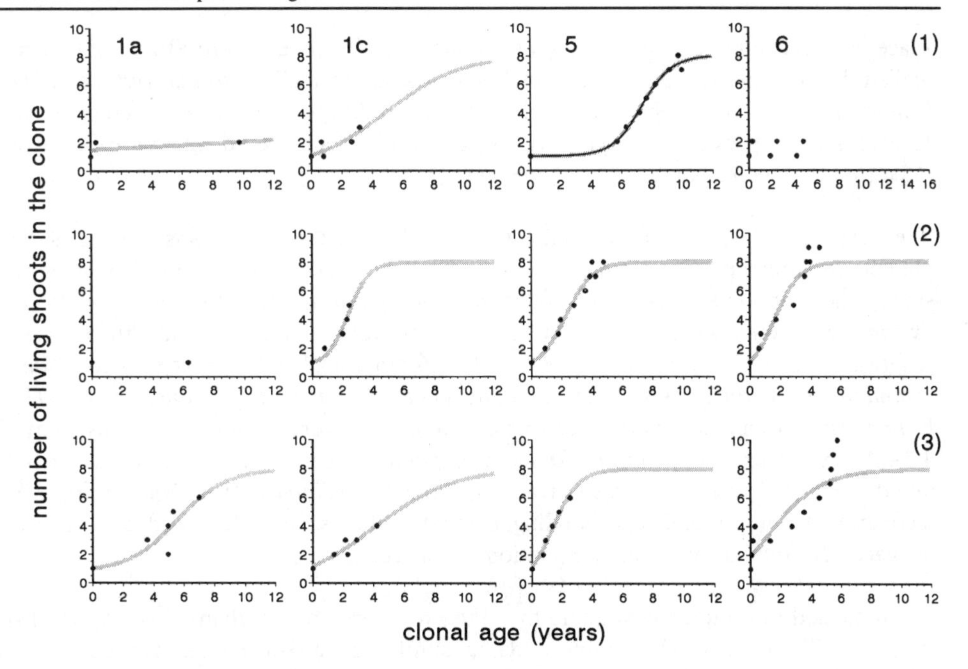

Fig. 4.7. Reconstructed net increase of the number of living shoots in an *Enhalus acoroides* clone. Three clones (1), (2) and (3) were collected each from sites 1a, 1c, 5 and 6. Curves are nonlinear fits (see Table 4.3 for regression statistics); no curves = data points are insufficient to fit such curves.

Table 4.3. Nonlinear regression statistics for the three *Enhalus acoroides* clones per site; r = intrinsic rate of vegetative propagation; $T_{4shoots}$ = time it takes from a single shoot to 4 shoots; x = value cannot be determined. The mean rhizome elongation rates ($cm.sht^{-1}.yr^{-1}$) are average values of the individual rates of all rhizome segments in a clone.

SITE	clone number	mean rhizome elongation rate ($cm.sht^{-1}.yr^{-1}$)	Nonlinear Regression R^2	r ($\%.yr^{-1}$)	$T_{4 shoots}$ (yrs.)
1a	(1)	4.07 ± 0.03	0.276	2.99	34.47
	(2)	1.90	x	x	x
	(3)	4.2 ± 1.74	0.760	61.85	5.56
1c	(1)	7.68 ± 1.21	0.705	41.75	4.94
	(2)	6.22 ± 1.66	0.931	143.63	2.37
	(3)	8.31 ± 1.40	0.881	39.85	4.07
5	(1)	4.35 ± 0.57	0.985	102.40	7.37
	(2)	4.67 ± 1.03	0.966	113.28	2.27
	(3)	4.15 ± 1.03	0.975	116.46	1.39
6	(1)	3.29 ± 0.83	x	x	x
	(2)	6.73 ± 1.05	0.904	120.87	1.64
	(3)	4.66 ± 0.99	0.782	566.78	1.05

space and reaching density levels of as high as 50 times more (Fig. 4.1) than in an undisturbed condition. Other species including the dominant *Enhalus acoroides* and *Thalassia hemprichii* are also somehow space-limited, being able to some extent recolonize within a relatively short period (ca. 2 yrs. post-disturbance) which would otherwise be difficult if, rather, either mechanism (1) or (2) operates.

The sequence of species establishment obtained from this study was consistent with the predictions from a previous demographic study (Vermaat et al. 1995) in the same area. In that study, the horizontal elongation rates for *Halophila ovalis* (141 cm apex^{-1} yr^{-1}) and *Syringodium isoetifolium* (134.7 cm apex^{-1} yr^{-1}) were found to be far higher than other seagrass species (33.9, 78.3, 5.3, 28.4 and 20.6 cm apex^{-1} yr^{-1} respectively for *Cymodocea rotundata*, *Cymodocea serrulata*, *Enhalus acoroides*, *Halodule uninervis* and *Thalassia hemprichii*). Comparable values for *Enhalus acoroides* were obtained from this study (Table 4.3). As can be expected, *Halophila* and *Syringodium* were the early recolonizers (first three months, Fig. 4.1; see also Preen et al. 1995 for *H. decipiens*, *H. spinulosa* and *H. ovalis* recovering from cyclone and flooding events). Other species followed and subsequently displaced *Halophila ovalis* in competition for space.

The projected times to achieve more than 90% recovery are less than 10 years for all species (Fig. 4.2; Table 4.1). This implies that a similar disturbance (e.g., typhoon 'blowouts', dugong-cleared patches) occurring every 10 years on the same spot may maintain the diversity of the mixed-species, shallow and clear-water meadows. However, this 'ideal' disturbance frequency (i.e., once every 10 years of 0.25 m^2 size) is probably not the same across different sites. As shown in this study, the recolonization by sexual propagules is less important even at a site (1a) where there is relatively high effort allocated to sexual reproduction (see Chapter 5 for *Enhalus acoroides*; pers. obs. for *Thalassia hemprichii*). Vegetative propagation may therefore be assumed as the main recolonization mechanism in most sites. It thus follows that the rate of recolonization mainly depends on the densities of the component species in the area surrounding the clearances (border effect). For the dominant species *Enhalus acoroides* and *Thalassia hemprichii*, shoot densities significantly vary between sites (Figs. 4.3, 4.4; Table 4.2). This suggests that recolonization of the cleared patches at the sites with low seagrass densities would be much slower (e.g., not a single recolonizer of either *Enhalus acoroides* or *Thalassia hemprichii* was observed in cleared gaps in sites 1c and 5 after 2 yrs. post-disturbance; data not shown).

The curve describing the increase in the number of living shoots in an *Enhalus acoroides* clone is necessary to approximate a crude relation among the recovery time, disturbance (clearance) size and the peripheral density. For instance at site 1a, taking the average shoot increase of *Enhalus* clones (Table 4.3), the "peripheral density" (number of *Enhalus* shoots along the 2-m perimeter of the clearance) should be about 2.5 shoots.m^{-1} to come up with such a recolonization curve for the species (Fig. 4.2). Increasing the disturbance size to, e.g., 1 m^2 (i.e., 100 cm x 100 cm), necessitates a "peripheral density" of ca. 5 shoots.m^{-1} to fully recover within 10 yrs. However, "peripheral density" is presumably constant within a site and hence the larger the disturbance size, the longer it takes to recover (see also Sousa 1984 for similar trends in various organisms in mussel beds). At the different sites where the intrinsic rate of increase in the number of living shoots of *Enhalus acoroides* can be much higher (e.g., sites 5 and 6, Table 4.3), only a single clone (i.e., 1 shoot along the 2-m perimeter) can fully recolonize a 0.25-m^2 clearance within 10 yrs. However at site 5, not a single *Enhalus* shoot was able to recolonize within ca. 2.5 years (data not shown), which suggests that the average "peripheral density" was much lower than 1.

For most species, the timing of disturbance did not affect the recolonization pattern (Fig. 4.1; Table 4.1). This means that, except for *Syringodium isoetifolium*, the multiplier of seagrass density (or specifically, the peripheral density) and rhizome elongation plus branching did not vary strongly over time (for contrasts, see Barrat-Segretain & Amoros 1995). This is supported by the virtual absence of temporal trends for *Enhalus* and *Thalassia* (Figs. 4.3, 4.4; Table 4.1). McManus et al. (1992) also observed that although seagrass biomass may vary annually, density is approximately constant. *Syringodium isoetifolium* is an exception as density becomes much higher in ca. July (pers. obs., 1993-1995). An important consideration which could lead to a strong temporal variation is desiccation. A study following the density of *Thalassia hemprichii* over time (Erftemeijer 1993) in a shallow (+0.3 m datum chart) site in Indonesia showed a significant temporal variation in density in coincidence with the occurrence of spring low tide during daylight. In this study, no similar 'burning' effect was observed since all the study plots were ≤ -0.5 m deep relative to chart datum. The lowest tide level observed during the study period was ca. -0.3 m.

To summarize, different species showed different strategies in recolonizing the artificially created gaps in a multi-species seagrass meadow. Relative to other coexisting species, the two structurally dominant species *Enhalus* and *Thalassia* were slow starters, and, although seedlings of these two species were able to colonize, they were much less important density wise. However, the recolonization curves of these two species had generally higher R^2-values (Table 4.1), implying much more steady recolonization trends. This may suggest that these two species are less influenced by the presence of other species, i.e., they are better competitors. Comparing *Thalassia* and *Enhalus*, the former colonized earlier and much faster than the latter. *Thalassia* fully recovered within 2 years after disturbance while *Enhalus* may achieve the same in about 10 years.

LITERATURE CITED

Barrat-Segretain, M.H. and C. Amoros. 1995. Influence of flood timing on the recovery of macrophytes in a former river channel. Hydrobiologia 316: 91-101.

Birch, W.R. and M. Birch. 1984. Succession and pattern of tropical intertidal seagrasses in Cockle Bay, Queensland, Australia. a decade of observations. Aquat. Bot. 19: 343-367.

Cambridge, M.L. 1975. Seagrasses of south-western Australia with special reference to the ecology of *Posidonia australis* Hook. *f.* in a polluted environment. Aquat. Bot. 1: 149-161.

Chiarello, E. and M. Barrat-Segretain. 1997. Recolonization of cleared patches by macrophytes: modelling with point processes and random mosaics. Ecological Modelling 96: 61-73.

De Iongh, H.H., B.J. Wenno and E. Meelis. 1995. Seagrass distribution and seasonal biomass changes in relation to dugong grazing in the Moluccas, East Indonesia. Aquat. Bot. 50: 1-19.

Duarte, C.M., N. Marba, N. Agawin, J. Cebrian, S. Enriquez, M.D. Fortes, M.E. Gallegos, M. Merino, B. Olesen, K. Sand-Jensen, J. Uri and J. Vermaat. 1994. Reconstruction of seagrass dynamics: age determinations and associated tools for the seagrass ecologist. Mar. Ecol. Prog. Ser. 107: 195-209.

Erftemeijer, P.L.A. 1993. Factors limiting growth and production of tropical seagrasses: nutrient dynamics in Indonesian seagrass beds. PhD Dissertation, Katholieke Universiteit Nijmegen, The Netherlands, 173 pp.

Fortes, M.D. 1988. Mangroves and seagrasses of East Asia: habitats under stress. Ambio 17: 207-213.

Holligan, P.M. and H. de Boois. 1993. Land-Ocean interaction in the coastal zone (LOICZ) - Science Plan. IGBP Global Change Report 25, ICSU-IGBP, Stockholm, 50 pp.

Holt, R.D., G.R. Robinson and M.S. Gaines. 1995. Vegetation dynamics in an experimentally fragmented landscape. Ecology 76: 1610-1624.

Hutchinson, G.E. 1978. An introduction to population ecology. Yale University Press, London, 260 pp.

McManus, J.W., C.L. Nañola, Jr., R.B. Reyes, Jr. and K.N. Keshner. 1992. Resource ecology of the Bolinao coral reef system. ICLARM Stud. Rev. 22, 117 pp.

Norusis, M.J. 1986. SPSS-PC+ manual. SPSS Inc. Chicago.

Paine, R.T. and S.A. Levin. 1981. Intertidal landscapes: disturbance and the dynamics of pattern. Ecol. Monogr. 51: 145-178.

Patriquin, D.G. 1975. "Migration" of blowouts in seagrass beds at Barbados and Carriacou, West Indies, and its ecological and geological implications. Aquat. Bot. 1: 163-189.

Poiner, I.R. 1989. Regional studies - Seagrasses of tropical Australia, In: A.W.D. Larkum, A.J. McComb and S.A. Shepherd (eds.). Biology of seagrasses, a treatise on the biology of seagrasses with special reference to the Australian region. Elsevier, Amsterdam, 279-303 pp.

Preen, A. 1995. Impacts of dugong foraging on seagrass habitats: observational and experimental evidence for cultivation grazing. Mar. Ecol. Prog. Ser. 124: 201-213.

Preen, A.R., W.J. Lee Long and R.G. Coles. 1995. Flood and cyclone loss, and partial recovery, of more than 1000 km^2 of seagrass in Hervey Bay, Queensland, Australia. Aquat. Bot. 52: 3-17.

Richards, F.J. 1969. The quantitative analysis of growth. In: F.C. Steward (ed.). Plant physiology - a treatise. Va. Analysis of growth: behavior of plants and their organs. Academic Press, London, pp. 3-76.

Rivera, P.C. 1997. Hydrodynamics, sediment transport and light extinction off Cape Bolinao, Philippines. PhD Dissertation, IHE-WAU, The Netherlands, 244 pp.

Short, F.T. and S. Wyllie-Echeverria. 1996. Natural and human-induced disturbance of seagrasses. Environmental Conservation 23: 17-27.

Sousa, W.P. 1984. Intertidal mosaics: patch size, propagule availability, and spatially variable patterns of succession. Ecology 65: 1918-1935.

Suchanek, T.H. 1983. Control of seagrass communities and sediment distribution by *Callianassa* (Crustacea, Thalassinidea) bioturbation. J. Mar. Res. 41: 281-298.

Thomas, L.P., D.R. Moore and R.C. Work. 1961. Effects of hurricane Donna on the turtle grass beds of Biscayne Bay, Florida. Bull. Mar. Sci. 11: 191-197.

Tilmant, J.T., R.W. Curry, R. Jones, A. Szmant, J.C. Zieman, M. Flora, M.B. Robblee, D. Smith, R.W. Snow and H. Wanless. 1994. Hurricane Andrew's effects on the marine resources. BioScience 44: 239-237.

Valentine, J.F., K.L. Heck, Jr., P. Harper and M. Beck. 1994. Effects of bioturbation in controlling turtlegrass (*Thalassia testudinum* Banks *ex* König) abundance: evidence from field enclosures and observations in the Northern Gulf of Mexico. J. Exp. Mar. Biol. 178: 181-192.

Vermaat, J.E., N.S.R. Agawin, C.M. Duarte, M.D. Fortes, N. Marba and J.S. Uri. 1995. Meadow maintenance, growth and productivity of a mixed Philippine seagrass bed. Mar. Ecol. Prog. Ser. 124: 215-225.

Vermaat, J.E. and M.J.M. Hootsmans. 1994. Intraspecific variation in *Potamogeton pectinatus* L.: a controlled laboratory experiment. In: W. van Vierssen, M.J.M. Hootsmans and J.E. Vermaat (eds.). Lake Veluwe, a macrophyte-dominated system under eutrophication stress. Geobotany 21, Kluwer Academic Publishers, Dordrecht, The Netherlands, 26-39 pp.

Williams, S.L. 1988. Disturbance and recovery of deep-water Caribbean seagrass bed. Mar. Ecol. Prog. Ser. 42: 63-71.

Chapter 5

Factors affecting the spatio-temporal variation in the sexual reproduction in *Enhalus acoroides* (L.f.) Royle

Abstract. Spatio-temporal differences in the flowering of *Enhalus acoroides* and the corresponding variation in environmental conditions were quantified. Although flowering was found to be a year-round phenomenon, flowering intensity varied temporally and appeared to correlate strongly with the variation in the mean water temperature. Spatially, the significant differences in flowering intensity correlated strongly with the differences in PAR availability at depth. Consistent with a previous study, flower scars on the rhizome of *Enhalus acoroides* in this study occurred in high frequency (ca. 4 times.shoot^{-1}.yr^{-1}). However, the frequency of flowers which actually mature was considerably less and varied spatially. This was clearly demonstrated by the *Enhalus acoroides* plants at the deepest site which rarely produce mature flowers while showing high frequency of flowers marks (and very young flowers) on their rhizomes. The spatial differences in the frequency of mature flowers were found to also correlate strongly with PAR. On the phenology of the release of male flowers, the influence of tides was found to be not significant, contrary to previous hypotheses. The most likely alternative mechanism is the production of gas bubbles, in which case, PAR availability probably plays a key role.

INTRODUCTION

Seagrasses are able to reproduce both sexually and asexually (Den Hartog 1970; Duarte & Sand-Jensen 1990). All species are able to produce flowers and fruits and also to produce new individuals through vegetative replication. While recently the information on the asexual means has been accumulating tremendously (e.g., Duarte & Sand-Jensen 1990; Gallegos et al. 1993; Duarte et al. 1994; Durako 1994; Marba et al. 1994; Vermaat et al. 1995), the relative importance of sexual reproduction for the establishment of meadows is, in general, poorly understood, except perhaps for some few species e.g., *Cymodocea nodosa* (Caye & Meinesz 1985; Terrados et al. 1993), *Thalassia testudinum* (Durako & Moffler 1987; Gallegos et al. 1992) and *Zostera* spp. (Hootsmans et al. 1987). This is particularly the case for *Enhalus acoroides*, a dominant, large, and widely-occurring species both locally in Bolinao (Meñez et al. 1983; Estacion & Fortes 1988; Vermaat et al. 1995; pers. obs., in deep and turbid sites) and regionally in the Indo Pacific coastal waters (Troll 1931; Johnstone 1979; Mwaiseje 1979; Brouns & Heijs 1986; Fig. 1.2, Chapter 1). Studies, so far, on the reproductive biology of *Enhalus acoroides* are scarce (Table 5.1) and dealt mostly with the anatomical details of the sexual reproductive structures (Troll 1931 and literature cited therein; Den Hartog 1970; Phillips & Meñez 1983; Pettit 1984).

Previous accounts quantifying spatio-temporal differences in the occurrence of flowers and fruits have been anecdotal. For instance, the prevailing belief that male flowering is related with spring tides (on the basis of fishermen's account that mass release of male flowers coincided with new- and fullmoon phases; Troll 1931) has not been tested. When released from spathes, male flower buds float like balloons (Svedelius 1904 in Troll 1931; pers. obs.) and transport is primarily dependent on wind velocity and/or surface currents (Troll 1931; Brouns & Heijs 1986). Discharge events are always remarkable because of the massive amount of white flower buds which, with the wind, move on the water surface as if racing

Table 5.1 Previous accounts of the reproductive biology of *Enhalus acoroides* (L.f.) Royle

Reference	Study Location	Results
Troll (1931)	Review (Ambon Malaysia)	- ripe fruits in September (Griffith 1851) - flowering period, quite limited: September to November (Zollinger 1854) - life-history of *Enhalus acoroides* (Svedelius 1904) - flower only when periodically emerged - detailed description of the morphology and development of female and male inflorescences (a) male inflorescences develop only at low tide (attributed to pressure and temperature) (b) ripe fruits from August to November (c) on pollination: clustering male flowers on the water surface; female flower stalks extend up to water surface; petals of male flowers, hydrophobic; petals of female flowers close when resubmerged (catching male flowers)
Johnstone (1979)	Papua New Guinea	- female flower peduncles (both from deep and shallow populations) grow exponentially - peduncle length is related to water depth where the plant is growing, i.e., longer in deeper sites
McMillan (1982)	Laboratory cultures	- *Enhalus acoroides* transplants failed to flower during five months under continuous light and temperatures - *Enhalus acoroides* did not flower during several years (1973-1982) of testing
Pettit (1984)	Review	- ripe pollen grain: spherical, trinucleate, inaperturate; 150-175 μm in diameter; grains per anther, 120 (Svedelius 1904), 29-98 (Cunnington 1909), 56 (Kausik 1941)
Brouns and Heijs (1986)	Papua New Guinea	- flowering, year-round phenomenon; male flower, hydrophobous and wind-driven - pollination occurred from March-August (as derived from from aged young fruits) during: (a) sufficiently low tides; (b) daytime; (c) new moon spring tidal levels, and; (d) SE moonsoon - fruit bursting occurred during spring-ebb tides (when ripe fruit is near water surface); fruit capsule disintegrates and decays within few days after opening - fruit development and ripening varied considerably (3-5 months) - 2-11 seeds per fruit; fruits with 3, 4 or 8 seeds were never observed - seeds (embryo) in initial stage of germination when the fruits opened; with loosely attached testa, may float in stagnant water for two days; without testa, will sink immediately
Uri et al. (Unpubl.)	Philippines	- on average, each flowering shoot had flowers during 25% of its life time (# flower scars per # internodes; n = 9) - 24 of the 42 shoots (~57%) never produced flowers; 5 of the 20 clones (25%) never produced flowers - per shoot basis, fruit weight is 3.3 times heavier than the average shoot weight (SW); a flower weighs 1/5 of SW. - annual fruit production calculated to be 8 m^{-2}; annual seed rain, 101 m^{-2} (together comprising 95% of the annual aboveground production)

against each other, and form large aggregates when trapped (Troll 1931; Brouns & Heijs 1986; pers. obs.). During discharge events in Bolinao, NW Philippines (Fig. 2.1, Chapter 2), male flower buds could be found in all parts of the reef flat area (pers. obs.). From these observations, it is apparent that male flowers are released about simultaneously, i.e., within a short period. Hence, it might be hypothesized that there exists a trigger (spring tide theory has been forwarded; Troll 1931) in the development and the subsequent release of male flowers from male spathes.

With regards to the spatial occurrence of *Enhalus* flowers, according to Troll (1931), *Enhalus* only flowers when occasionally emerged. Hence, no flowers may be found in deep, permanently submerged meadows. Practically no quantitative evidence exists which confirms or denies Troll's observations. Although such a rigorous depth limitation seems quite logical for female *Enhalus*, which should extend its flower peduncle to reach the water surface (because the capture of the hydrophobous male flowers is only possible at the water surface), this might not be true for males which may discharge the tiny male flowers underwater. Based on the findings in the earlier chapter (Chapter 3), where deep populations were estimated to have little photosynthetic resources for anything but leaf production, some support for depth limitation to flowering exists, although this is linked to energy availability and not tidal emergence.

On temporal aspects, there are two recent quantitative studies (Brouns & Heijs 1986 and Uri et al. Unpublished; Table 5.1). The former quantified the flowering of *Enhalus* in permanent plots within one site and the latter presented results obtained by using an age reconstruction technique (Duarte et al. 1994; Gallegos et al. 1992). Brouns and Heijs (1986) showed that *Enhalus* flowering is a year-round phenomenon. Applying the reconstruction technique, Uri et al. (Unpublished; Table 5.1) showed that the frequency of flower scars on *Enhalus* rhizomes is four times.yr^{-1} and, based on this frequency, estimated an annual seed rain of 101 seeds.m^{-2}. This value may suggest a substantial contribution of sexual reproduction in the establishment and maintenance of *Enhalus*-dominated meadows. Considering an establishment success ca. 37% for *Enhalus* seedlings at the site (based on the data for site 1a; Chapter 6), there should approximately be about 37 seedlings.m^{-2}.yr^{-1}. However, the records on the settlement success of *Enhalus* seedlings on the disturbed and undisturbed plots in a shallow, clear water meadow in about the same location (Bolinao, Philippines; Chapter 4) showed only 0.67 seedling.m^{-2}.yr^{-1}. Apparently, either there exists a big loss factor which has not been considered (e.g., seed and/or fruit export factor) or, one or more of the terms (e.g., flowering percentage and frequency, fruiting success, seed number per fruit) in the extrapolation steps have been overestimated. A rigorous comparison of true observations and reconstruction extrapolation seems warranted.

In view of the above, this study has the following objectives: (1) to test whether *Enhalus* flowering differs temporally and spatially, and if possible, to correlate these differences with the corresponding environmental variables especially light and tides (see also Chapter 3); (2) to test whether the mass release of male flowers is correlated with the tidal cycle; and (3) to compare the frequency of mature flowers (observed) with that of flower scars (age reconstruction technique).

MATERIALS AND METHODS

Spatio-temporal variation in the intensity of flowering and fruiting

Flowering and fruiting of *Enhalus acoroides* were quantified in sites 1a, 1c, 4, 5 and 6 (see Fig. 2.1, Chapter 2) from April 1994 to February 1996. During each sampling, ca. 25 quadrats (50 x 50 cm) were randomly thrown, until at least 100 *Enhalus* shoots had been examined. Each *Enhalus* shoot within each quadrat was checked for flower occurrence. For female flowers (♀), four stages were distinguished: 1 = emerging, young flower; 2 = mature flower (about ready to be pollinated; Fig. 1.1, *Enhalus*, letter "h", Chapter 1); 3 = rotting, unfertilized flower; and 4 = mature fruit (Fig. 1.1; *Enhalus*, letter "i", Chapter 1). For male inflorescence (♂; Fig. 1.1, *Enhalus*, letter "d"), two stages were distinguished: 1 = male spathe still full with flower buds, and; 2 = male flower buds had been released. Intensity of these stages was expressed as percentages of the total number of shoots counted.

Spatial variation in flowering frequency

Tagged shoots
To determine the number of times annually that an average shoot actually flowers, ten shoots of *Enhalus acoroides* at each of the sites 1a, 1b, 1c and 5, were tagged and checked (once per 1-2 mos.) for flowering and fruiting using the same distinction as above. When actually produced, inflorescence structures may remain with the plant for at least 2 months.

Flower scar-based quantification: ageing technique
Similar to leaf scars, flower marks can be found on the rhizomes (see Fig. 1.3, Chapter 1). Thus, flowering frequency may also be quantified by age reconstruction of rhizome samples (Duarte et al. 1994; Gallegos et al. 1992). For this purpose, three clones of *Enhalus acoroides* were collected from each of the sites 1a, 1c, 5 and 6. All rhizome segments were aged using plastochrone interval (PI) values previously determined (cf. Chapter 4). Flowering frequency of a rhizome segment (see also Fig. 1.3 A & B, Chapter 1) was basically calculated as the number of flower marks in the segment divided by the time elapsed to attain the total segment length. Segment age was calculated as the sum of all the PIs (converted to actual time) belonging to the same rhizome axis, i.e., including all other previous segments starting from branching point (or seed, or when cut, cluster start) up to the last PI of the segment (see also Fig 1.3A, Chapter 1).

Phenology of male flower discharges

To test whether the timing of the male flower release correlated with one or more of the tidal characteristics, all observed male flower discharges were recorded. On each day when a discharge was observed, the corresponding tidal characteristics (type, levels, rising/falling, etc.) were obtained.

Seed number per fruit

To quantify the variability in the number of seeds per fruit of *Enhalus acoroides*, a total of 190 *Enhalus* fruits were harvested, opened and seeds per fruit were counted. Fruits were collected in July 1994, November 1994 and December 1995 from a wider area but with conditions similar to site 1a (datum depth ca 0.5 m; multi-species meadow). The mean number of seeds obtained from these separate collections did not significantly vary (Tukey test, $p > 0.05$). Samples were then pooled to obtain an overall mean.

Relevant factors affecting sexual reproduction

PAR, temperature, tides and exposure duration
Calculations of PAR at seagrass depth were done by integrating daily PAR from sunrise to sunset, correcting for cloud/rain attenuation, datum depth ± tide level, and the mean water extinction coefficient of the site (see also Chapter 3). For temperature and tides, this chapter uses the appropriate data sets presented in Chapter 2. Exposure duration of stage ♀2 flower tips (which are ca. equal to the height of the leaf tips; pers. obs.) was calculated from the sinusoidal tidal curve of the day and the measured datum depths of flower tips. Assuming that the release of male flowers occurs during daytime (Brouns & Heijs 1986; pers. obs.), the daytime exposure was separated from the nighttime exposure.

Biomass and nutrient costs of producing sexual reproductive structures
To check the possibility that spatial differences in flowering might be due to carbon or nutrient (N, P) limitation, the cost of producing sexual reproductive structures was quantified in terms of units C, N or P. At site 1a where flowering plants were abundant, 30 shoots with reproductive structures (10 with male ♂1 inflorescence, 10 with female ♀2 flowers and 10 with fruits, ♀4) were gathered. Reproductive structures included inflorescence peduncle, spathe, flowers, pericarp and seeds. As "non-reproductive" baseline, ten shoots each of *Enhalus acoroides* without reproductive structures were randomly collected at sites 1a, 1b, 1c, 4, 5 and 6. After washing-off periphyton and attached sediment and separating leaves, roots, rhizomes (ca. 6-PI size equivalent) and reproductive structures, samples were oven-dried at 70°C until weights were constant (24-48 hrs).

Nutrient analyses of the plant samples were done at the Institute for Infrastructural, Hydraulic and Environmental Engineering (IHE, Delft, The Netherlands; see also Knight 1996). To determine total nitrogen and phosphorus, plant samples, after powdering using mortar and pestle, were digested with H_2SO_4/Se/Salicylic acid and H_2O_2 (according to Kruis 1995). Ash-free dry weight (AFDW) was determined by combustion at 550°C for six hours. Organic carbon was assumed to be 40% of AFDW (Hootsmans 1994).

Values are expressed on per shoot basis, i.e., one shoot is composed of ca. 5-6 leaves and the corresponding amount (6 PIs) of rhizome internodes and associated roots.

Data analyses

After data transformations to control normality and homogeneity of variance, spatio-temporal differences in various parameters were tested using ANOVA. When mutiple comparisons were necessary, the Tukey test was used. Differences were considered significant when $p < 0.05$. Correlations of two variables (e.g., temperature vs. flowering; PAR vs. flowering) were tested by linear regression. For the temperature-flowering tests, only those variable-pairs having significant correlations are shown.

Table 5.2. Two-way ANOVA testing the effects of site and sampling month on the abundance of *Enhalus acoroides* sexual reproductive structures quantified using random quadrats; n = 25 (quadrats).

	p-value	variance component
total female flowers		
site	< 0.001	8.32%
sampling month	< 0.001	1.71%
interaction	< 0.001	1.70%
basic error (i.e., between quadrats)		88.27%
fruiting		
site	< 0.001	5.80%
sampling month	< 0.001	1.52%
interaction	< 0.001	9.46%
basic error		83.22%
total male flowers		
site	< 0.001	2.77%
sampling month	< 0.001	1.63%
interaction	< 0.001	9.17%
basic error		86.42%
total flowering		
site	< 0.001	10.78%
sampling month	< 0.001	3.15%
interaction	< 0.001	6.13%
basic error		79.94%

RESULTS

Spatio-temporal variation in the intensity of flowering and fruiting

All main and interactive effects of sampling month and site factors on the intensity of *Enhalus* flowering were significant (Table 5.2). Comparatively, the variance component

Fig. 5.1 (opposite page, left side). Female flowering intensity at different sites using random quadrats: total = different flower categories were not distinguished (i.e., first 3 sampling months A, M and J); for flower categories, see materials and methods; the mean water temperature curve is shown for sites 1a and 4 but in fact in the same curve also prevails at other sites.

Fig. 5.2 (opposite page, right side). Male flowering intensity at different sites using random quadrats: total = different stages not distinguised; other legends = see materials and methods.

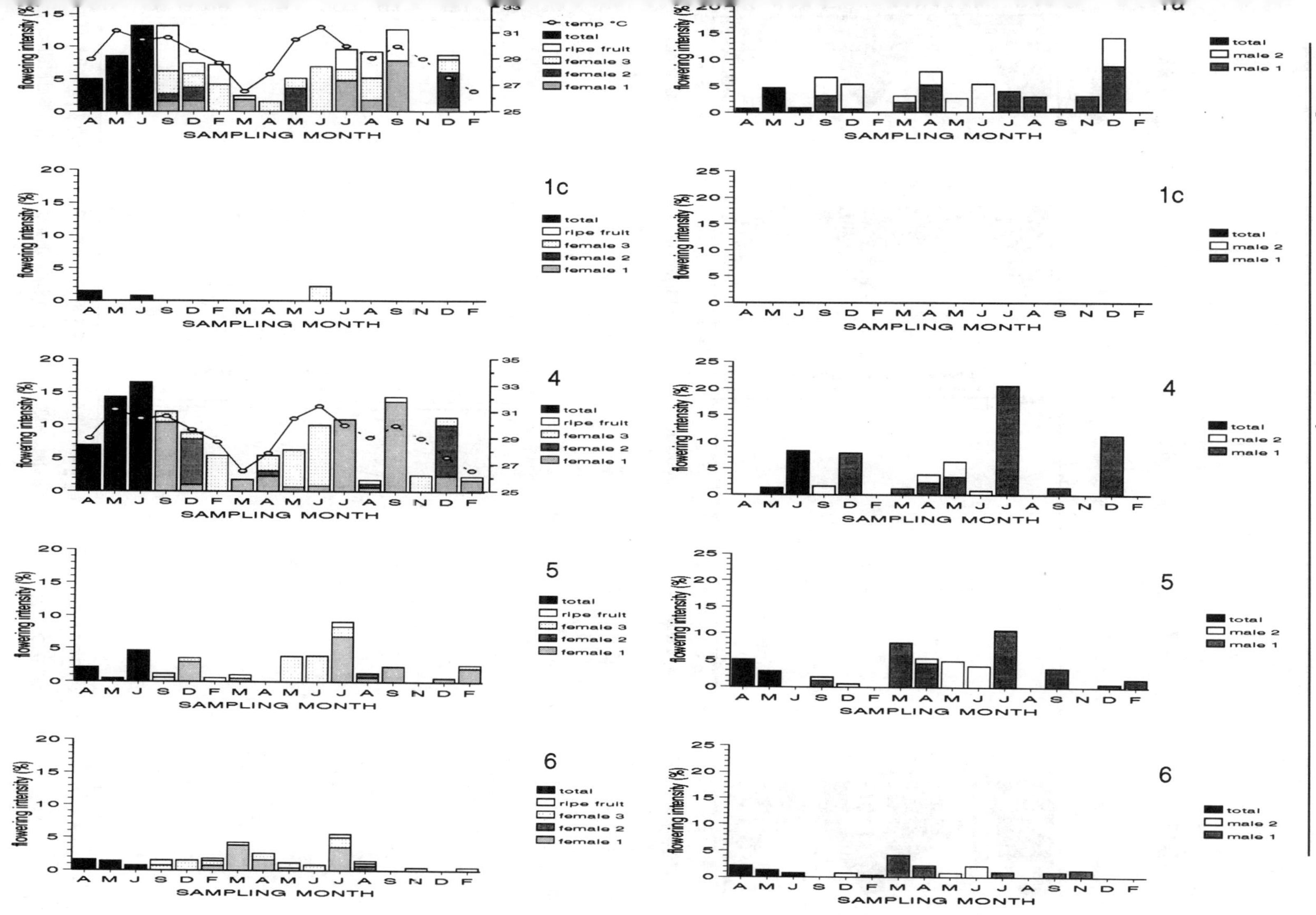
flowering intensity (%)
SAMPLING MONTH
A M J S D F M A M J J A S N D F
1c
4
5
6
temp °C
total
ripe fruit
female 3
female 2
female 1
male 2
male 1

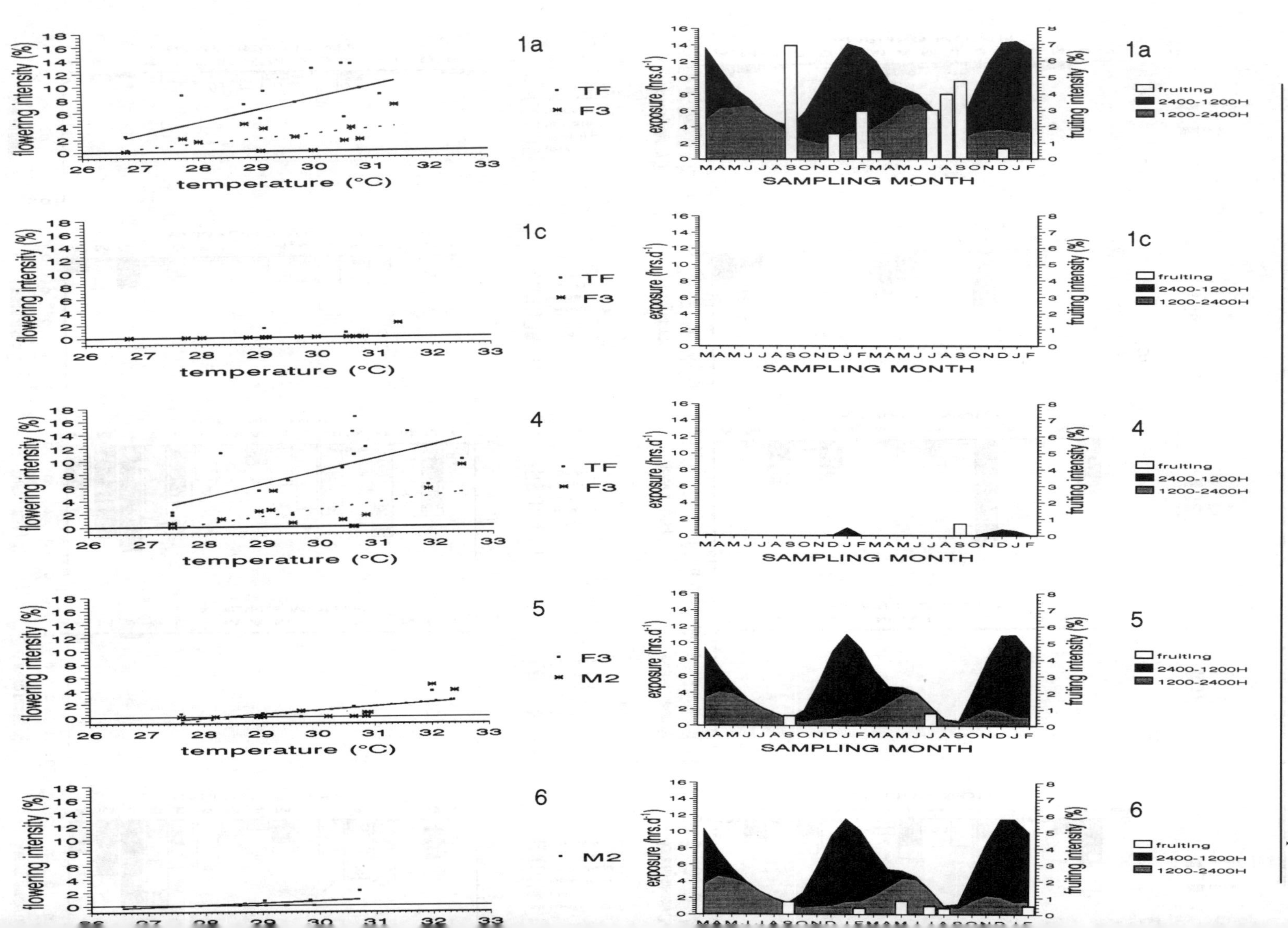
1a
TF
F3
flowering intensity (%)
temperature (°C)
exposure (hrs.d-1)
fruiting intensity (%)
SAMPLING MONTH
fruiting
2400-1200H
1200-2400H
1c
4
5
M2
6

accounted by the site factor was, in all cases, greater than the component due to the differences in sampling time although both were relatively low. The interactive effects of both factors appeared strongest in the abundance of male flowers and fruits (Table 5.2).

Temporally, the trend of the variation in the intensity of female flowering at sites 1a and 4 was most pronounced (Fig. 5.1). This trend correlated significantly with water temperature (Figs. 5.1, 5.3). The temporal variation in the intensity of male flowering had no clear trend (Figs. 5.2) except that at sites 5 and 6, the correlations with water temperature were also significant (Figs. 5.2, 5.3). The occurrence of fruits, which is a product of the synchrony of the male and female flowers and exposure (low water to allow pollination), had also no clear annual trend (Fig. 5.4), although at site 1a, the high number of fruits observed during July 1995 - September 1995 period was probably a result of the long exposure duration at daytime during April 1995 - June 1995. This suggests a 3-month lag-time between pollination and ripening.

Across sites, differences were clear (Table 5.3). For both male and female flowering, values found at the shallow- and clear-water site (1a) were higher than those found at the deepest- and clear-water site 1c (Table 5.3). In general, *Enhalus* flowering intensities at the darker environments (deep and/or turbid) were lower and the general trend strongly correlated with the mean PAR availability (Fig. 5.5) which explained convincingly the rarity of flowering at the darkest environment (site 1c). This was supported by the finding that the rare flowers at site 1c occurred around the periods when the mean PAR available was more than 12 $E.m^{-2}.d^{-1}$ (Fig. 5.6).

The abundance of fruits also clearly differed between sites (Table 5.3; Fig. 5.4). The influence of exposure duration on fruit formation was most obvious when comparing sites 1a (shallow; 0.35 m) and 4 (deep; 1.2 m) which, although had similar female flowering intensity values (Table 5.3; Figs. 5.1, 5.2, 5.3), differed significantly in fruit abundance (Fig. 5.4; Table 5.3).

Spatial variation in frequency of flowering

Tagged shoots

An average *Enhalus* shoot flowered ca. once a year (1.05 ± 0.24 $flowers.yr^{-1}$, 1a; 1.29 ± 0.22 $flowers.yr^{-1}$, 1b; Table 5.4) at sites 1a and 1b, significantly more frequent than at site 1c (0.10 ± 0.10 $flowers.yr^{-1}$; Table 5.4). The mean flowering frequency of *Enhalus* at site 5 (0.73 ± 0.34 $flower.yr^{-1}$) was an intermediate level. At this site, only five of ten tagged shoots were able to produce flowers during the entire period and these were all males. However, relative to other sites, the variation between shoots at site 5 was high, with one shoot producing 3.43 $flowers.yr^{-1}$ (Table 5.4). Similar to flowering intensity (Fig. 5.5), the correlation of flowering

Fig. 5.3 (opposite page, left side). Flowering intensity regressed with water temperature. Except 1c, shown lines are only those which are significant at $p < 0.05$. TF = total flowering; F3 = female 3; M2 = male 2 (see Materials and Methods).

Fig. 5.4 (opposite page, right side). Fruiting intensity (bars) plotted with the probable exposure duration of the female flowers (area graph). Different area shadings represent time period within the 24-hr day. Total area graph height represents the total exposure per day.

Table 5.3 Mean intensity (% of shoots counted) of *Enhalus acoroides* flowering at the study sites. Different letters (superscript) attached to values indicate significant (Tukey test, $p < 0.05$) differences across sites. ns = not significant; nv = no value; female: 1 = young, emerging; 2 = ready for pollination; 3 = rotting, unfertilized; 4 = fruit; male spathe: 1 = full of flower buds; 2 = empty spathe.

	flower category	Site 1a	1c	4	5	6
Female	1	1.52 ± 0.64 ^{b}	0.00 ± 0.00 ^{a}	3.54 ± 1.34 ^{c}	1.22 ± 0.54 ^{ab}	0.75 ± 0.38 ^{ab}
	2	0.93 ± 0.47 ^{b}	0.00 ± 0.00 ^{a}	1.22 ± 0.75 ^{b}	0.05 ± 0.05 ^{a}	0.06 ± 0.06 ^{a}
	3	2.06 ± 0.56 ^{bc}	0.16 ± 0.16 ^{a}	2.29 ± 0.77 ^{c}	0.92 ± 0.38 ^{ab}	0.62 ± 0.14 ^{a}
	4	1.88 ± 0.63 ^{b}	0.00 ± 000 ^{a}	0.06 ± 0.06 ^{a}	0.10 ± 0.07 ^{a}	0.25 ± 0.09 ^{a}
	total	6.85 ± 1.08 ^{c}	0.27 ± 0.16 ^{a}	8.13 ± 1.21 ^{c}	2.31 ± 0.58 ^{b}	1.60 ± 0.37 ^{ab}
Male	1	2.25 ± 0.71 ^{b}	0.00 ± 0.00 ^{a}	3.68 ± 1.70 ^{c}	2.30 ± 0.97 ^{c}	0.76 ± 0.34 ^{ab}
	2	1.92 ± 0.60 ^{c}	0.00 ± 0.00 ^{a}	0.52 ± 0.52 ^{ab}	0.82 ± 0.44 ^{b}	0.32 ± 0.18 ^{ab}
	total	3.78 ± 0.89 ^{b}	0.00 ± 0.00 ^{a}	4.02 ± 1.41 ^{b}	3.03 ± 0.79 ^{b}	1.15 ± 0.28 ^{a}
Total		10.64 ± 1.46 ^{d}	0.27 ± 0.16 ^{a}	12.15 ± 2.19 ^{d}	5.35 ± 1.19 ^{c}	2.74 ± 0.56 ^{b}
Female-Male ratio		1.81	nv	2.02	0.76	1.39

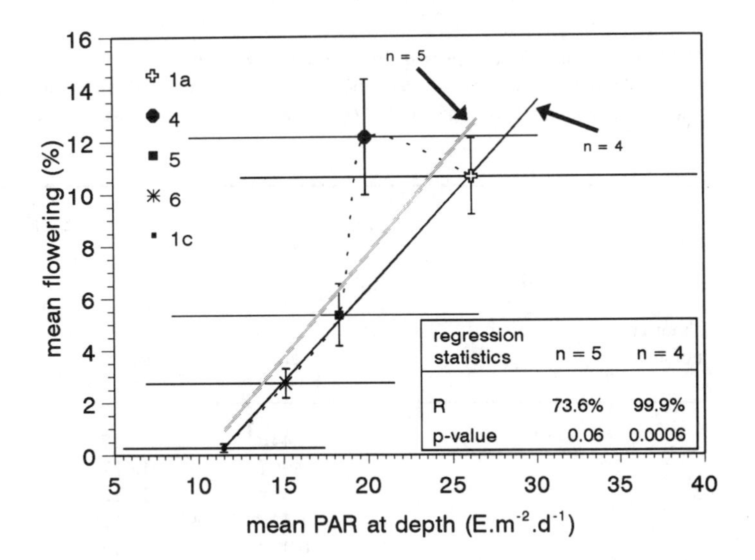

Fig. 5.5. Mean total (female + male) flowering intensity of *Enhalus acoroides* at the sites plotted against the mean PAR at seagrass depth at the respective sites. Vertical lines are standard errors of flowering. Horizontal lines indicate minimum and maximum PAR values. Two regression lines are shown: n = 5 includes all the five sites; n = 4 disregards site 4.

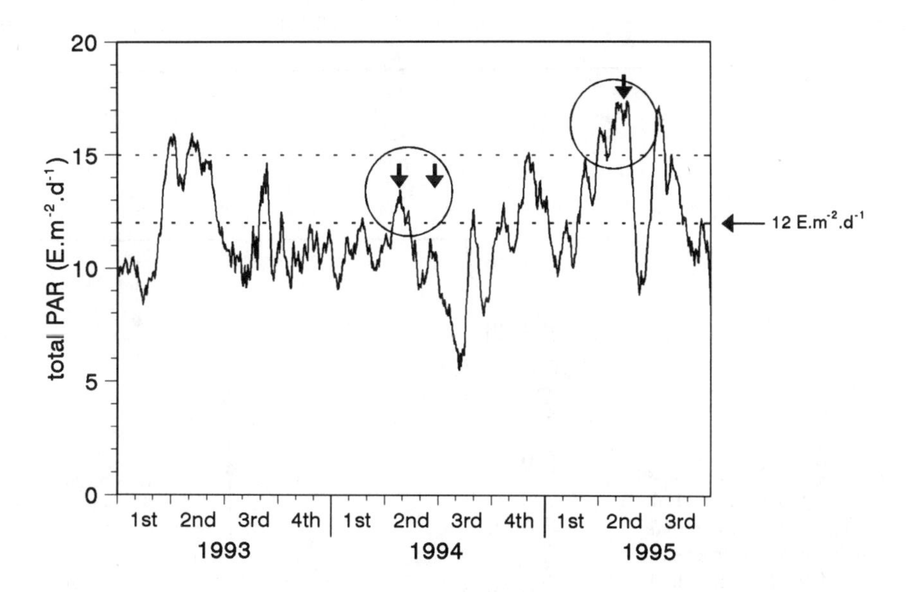

Fig. 5.6. Mean PAR available at site 1c. Curve was smoothened by taking 15-day running averages. The dates when rare mature flowers were observed are marked with arrows. These rare flowering events (encircled) at the site tend to occur when mean PAR values were relatively higher, i.e., more than 12 E.m^{-2}.d^{-1}.

Table 5.4. Annual flowering frequency of the tagged *Enhalus acoroides*; ♂ = male plant; ♀ = female plant; ? = sex unknown; x = died; n.v. = frequency could not be estimated; different letters (superscript) attached to mean values indicate significant differences ($p < 0.05$, Tukey test) between sites.

		SITE			
		1a	1b	1c	5
	depth, m	0.5	1.25	3.0	0.6
	K_d, m^{-1}	0.3	0.3	0.3	0.6
shoot tag number	sediment type	sandy	sandy	sandy	muddy
1		1.44♂	1.52♀	0.00?	3.43♂
2		0.96♀	1.01♀	0.00?	0.91♂
3		2.89♀	1.52♂	0.00?	0.00?
4		1.44♂	0.51♂	0.00?	0.51♂
5		0.49♀	2.53♂	0.00?	1.52♂
6		0.48♂	1.01♀	0.00?	0.00?
7		0.96♂	2.02♀	0.00?	0.00?
8		0.00?x	x; n.v.	0.00?	0.00?
9		0.92♂	1.01♀	0.00?	0.00?
10		0.96♂	0.51♂	1.01♀	0.91♂
mean		1.054^{b}	1.293^{b}	0.101^{a}	0.728^{ab}
se		0.245	0.224	0.101	0.345

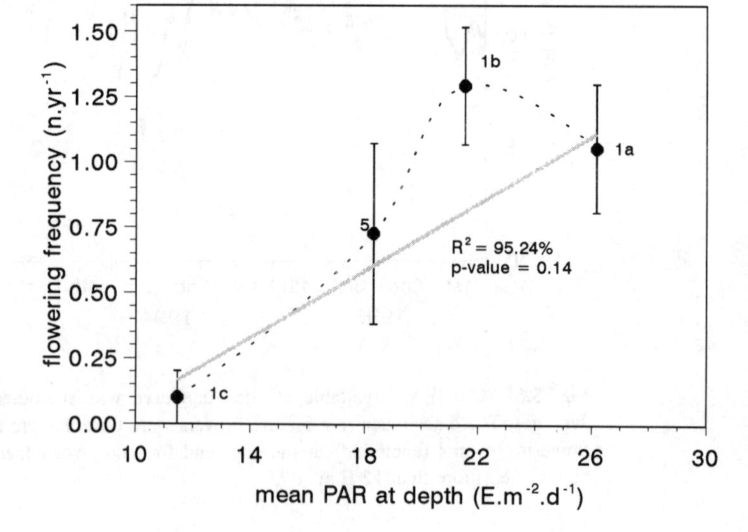

Fig. 5.7. Mean annual flowering frequency of the 10 tagged *Enhalus acoroides* shoots randomly selected at each of the sites 1a, 1b, 1c and 5. As in Fig. 5.5, value for 1b was disregarded for the regression showing high R^2 (95%) but high p-value due to small n.

Table 5.5. Frequency of flower scars on the rhizome of *Enhalus acoroides* (L.f.) Royle and the number of developed flowers found in between leaf bases. Different letters (superscript) attached to values indicate significant differences (Tukey test, $p < 0.05$) between (1) clones per site and (2) sites. ns = not significant; standing shoots = living shoots at the time of harvest; developed flowers = very young or otherwise distintegrating (?), developed flowers found in between leaf bases.

Site	Clone	Frequency ± se (n flower scars per year)	Standing shoots (n)	Number of developed flowers	Mean n developed flower per standing shoot
1a	1	4.81 ± 0.62^{ns}	2	1	0.50
	2	6.32	1	0	0.00
	3	2.95 ± 0.60^{ns}	6	13	2.17
1c	1	6.27 ± 0.16^{b}	3	8	2.67
	2	1.76 ± 0.84^{a}	5	2	0.40
	3	6.05 ± 0.28^{b}	4	9	2.25
5	1	5.49 ± 0.50^{b}	7	16	2.29
	2	1.01 ± 0.21^{a}	8	20	2.50
	3	6.25 ± 0.32^{b}	6	11	1.83
6	1	1.37 ± 0.77^{a}	2	0	0.00
	2	4.51 ± 0.48^{b}	9	17	1.89
	3	3.78 ± 0.64^{ab}	11	20	1.82
Overall					
1a		3.55 ± 0.52^{ns}	9	14	1.52
1c		4.50 ± 0.51^{ns}	12	19 ?	1.58
5		3.92 ± 0.42^{ns}	21	47	2.24
6		3.72 ± 0.40^{ns}	22	37	1.68

frequency with PAR availability was suggestive (R^2 = 95%; Fig. 5.7) although not significant at $p < 0.05$ mainly due to low sample size (n = 3, excluding site 1b).

Flower scar-based quantification: ageing technique
Differences between clones collected from the same sites were significant but, in general, the frequency of flower scars on the rhizomes of *Enhalus acoroides* was high for all of the *Enhalus* clones collected. An average *Enhalus* shoot produced flower scars ca. 4 times annually and differences between sites were not significant (Table 5.5) which is in strong contrast with the results from tagged shoots. Shoot age had no clear influence on flowering frequency (Fig. 5.8). What was most remarkable was the high frequency (4.5 ± 0.51 flower scars.yr^{-1}; Table 5.5) of flower scars on the rhizomes of *Enhalus* at site 1c where flowers were rarely observed (Tables 5.3, 5.5; Figs. 5.1, 5.2). In between the leaves of standing shoots, the frequency of occurrence of mainly disintegrating "flowerlets" (smaller than stage ♀1 or ♂1) was also high at site 1c (Table 5.5). Apparently, flower primordia were produced regularly, but the chance that those develop further is small and probably depends largely on light availability.

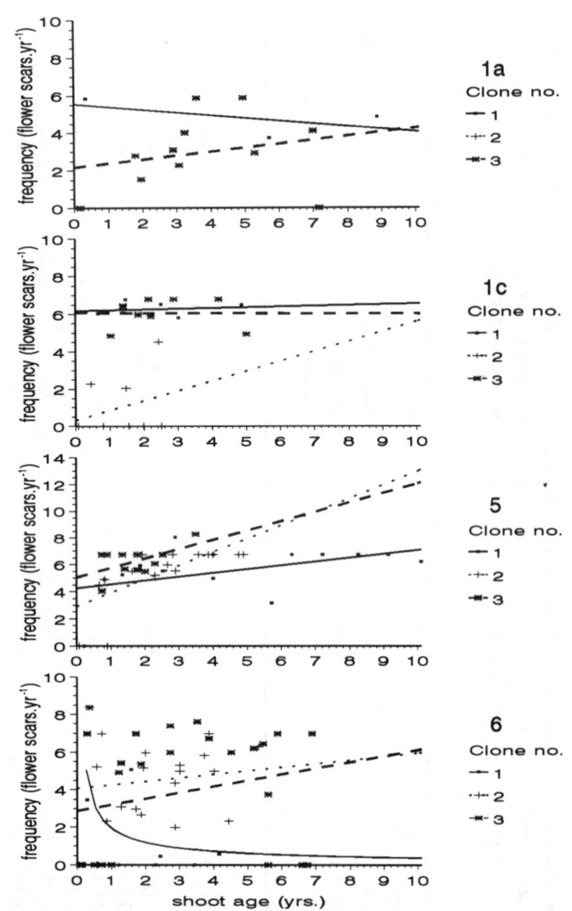

Fig. 5.8. Frequency of flower scars on *Enhalus* rhizomes (age reconstruction technique) plotted against shoot age. Shoot age, in this case, was calculated as the cummulative age from the branching point, (Fig. 1.3, Chapter 1) or 0,0 x-y coordinates in the case of the main axis (first shoot). Curves are placed to show general trends in flowering frequency and shoot age relations per clone.

Phenology of male flower discharges

The synchronous release of male flowers was observed all year-round (Fig. 5.9). There was no clear pattern of the occurrence of heavy (i.e., reef flat wide and in large aggregates of male flowers) and light (few, sparse) discharge events. From these observations (Fig. 5.9), it was shown that:

(1) mass releases of male flowers happened during both spring (high amplitude) and neap (low amplitude) tides ; they also occurred in all moon phases (not illustrated here);
(2) there was no strong influence of water level during the approximate release period;
(3) mass releases were not dependent on the minimum water level of the day during the event; and
(4) mass releases occurred during both rising and falling tide levels.

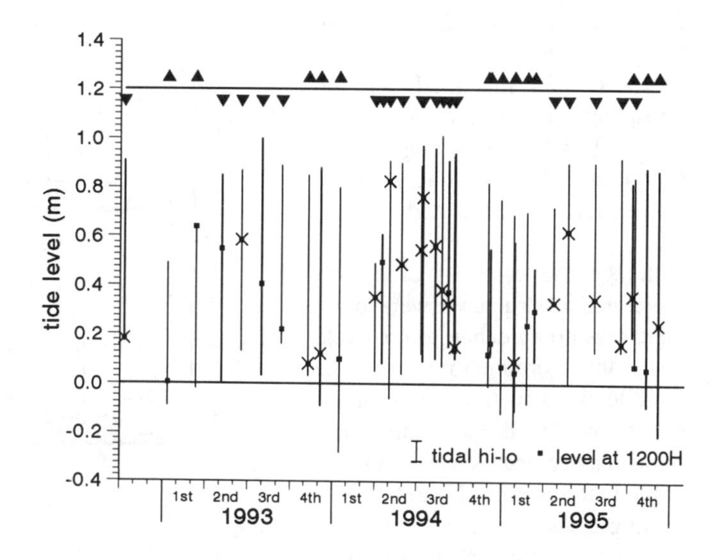

Fig. 5.9. Discharges of *Enhalus acoroides* male flowers. Vertical lines indicate the high and low tide levels of the day coinciding with each discharge event. The levels of the symbols (▪ = light discharge; ✕ = heavy discharge) with respect to the y-axis represent the tide level at noontime (the approximate time when discharge commences) on the day of the discharge. Triangle above each event indicates whether, at noontime, the tide was rising (▴) or falling (▾).

Seed number per fruit

The frequency distribution of the number of seeds per fruit was approximately normal (bell-shaped), ranging from 3-15 with a mean ± sd of 9.09 ± 2.49 (n = 190, Fig. 5.10). This result compliments earlier figures reported by Den Hartog (1970; range, 8-14 seeds per fruit) and Brouns and Heijs (1986; range, 2-11 seeds per fruit). The latter did not find fruits with 3, 4 or 8 seeds.

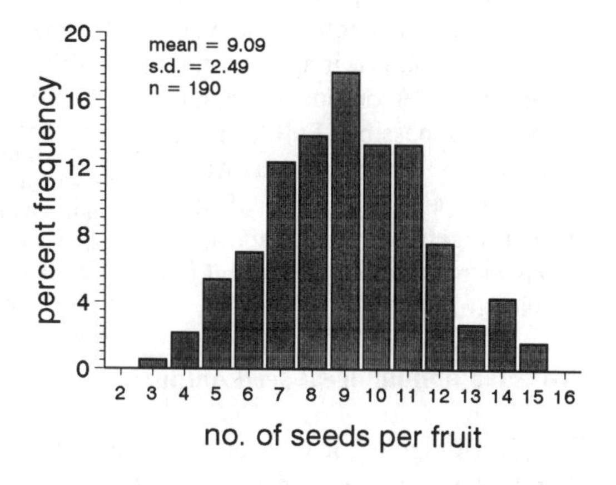

Fig. 5.10. Percent frequency distribution of the number of seeds per fruit of *Enhalus acoroides*.

Biomass and nutrient costs of producing sexual reproductive structures

The non-reproductive structures (leaves, roots and rhizomes) of the female reproductive shoots tended to be heavier ($p < 0.05$) than those of the corresponding non-reproductive plants (♀2/♀4 vs. 1a; Fig. 5.11). The mean total weight of the non-reproductive structures in shoots with male inflorescences was not different from that of the non-reproductive shoots. However, the cost of producing a female flower (567.13 ± 95.75 mg DW) was higher than producing a male flower (127.92 ± 15.16 mgDW). When a female flower would eventually develop into a fruit, the cost would be almost 10-fold more in magnitude (4890.46 ± 296.01 mgDW). In parallel amounts (Fig.

5.11), a shoot needed nitrogen (3.84 ± 0.26, 14.76 ± 1.13 and 63.19 ± 4.86 mg total N) and phosphorus (0.77 ± 0.26, 2.27 ± 0.57 and 14.58 ± 0.49 mg total P) to produce, respec-tively, male flower, female flower and fruit.

Although the costs of producing reproductive structures were high, there was no evidence to conclude that the production of the reproductive structures of the larger plants at the deeper and more turbid environments were nutrient limited. If the total N and P in a fruiting *Enhalus* at site 1a were to be used as maximum available values (ca. 100 mg total N and 22 mg total P, respectively), then there should still be enough nutrients to produce, at least, male (♂1) or female (♀2) flowers at, for instance, site 1c (Fig. 5.11), but the allocation is apparently to leaves. Also, in all stations, it appears that the male flower production ought to have been possible. Full fruiting, however, appears almost impossible also for most of the smaller vegetative shoots at 1a, if resources have to be reallocated to produce fruits.

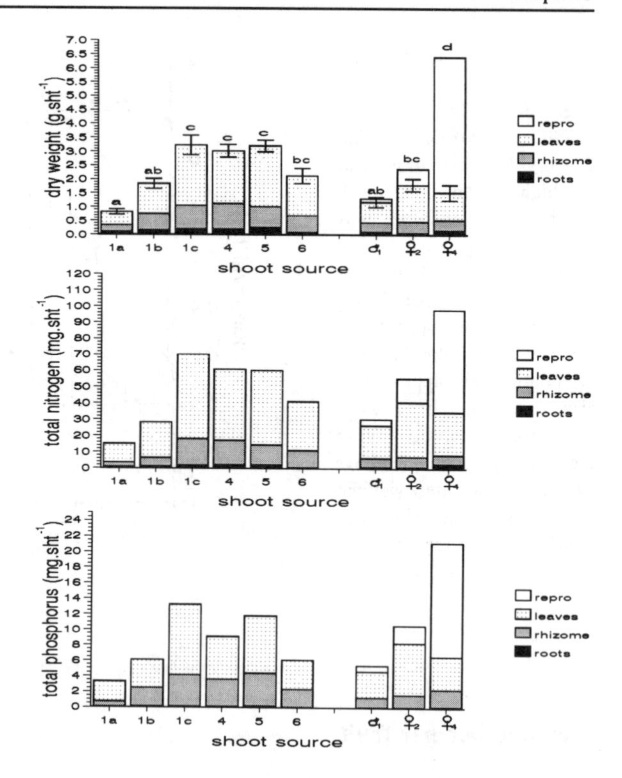

Fig. 5.11. Allocation of biomass (organic carbon is ca. 40% AFDW), nitrogen and phosphorus in non-reproductive *Enhalus acoroides* at the different sites and reproductive shoots at site 1a. ♀2 = 1a shoot with stage 2 female flower; ♀4 = 1a shoot with mature fruit; ♂ = 1a shoot with male 1 flower; repro = reproductive stuctures. Different letters above bars indicate significant differences at $p < 0.05$ (Tukey test). Not indicated: 1a total shoot mean biomass is significantly lower than either ♀2 (excluding repro) or ♀4 (excluding repro); differences in nutrient (N and P) allocations which are parallel with those of carbon. In all cases, n = 10.

Projected annual areal seed output

From the foregoing, it was found at site 1a that:

(1) *Enhalus acoroides* density is ca. 10 shoots per quadrat (1 quadrat = 0.25 m^2; Chapter 4);
(2) an average shoot flowers 1.054.yr^{-1} (Table 5.4);
(3) the fruiting-female flower ratio is ca. 1.88 out of 4.79 females (6.85-2.06; Table 5.3);
(4) the average fruit contains ca. 9.09 seeds (Fig. 5.10);
(5) seedling establishment at 1a is ca. 37% (Chapter 6).

Then, the gross areal seed output at site 1a may therefore be calculated as:

$$n\ seeds.quadrat^{-1}.yr^{-1} = \frac{10\ shoots}{quadrat} * \frac{1.054\ flowers}{shoot * yr} * \frac{6.85\ females}{10.64\ flowers} * \frac{1.88\ fruits}{4.79\ females} * \frac{9.09\ seeds}{fruit}$$

$$=> 23\ seeds.quadrat^{-1}.yr^{-1} \quad => 9\ seedlings.quadrat^{-1}.yr^{-1}$$

$$=> 93\ seeds.m^{-2}.yr^{-1} \quad => 35\ seedlings.m^{-2}.yr^{-1}$$

This gross areal seed output (93 seeds.m^{-2}.yr^{-1}) appears to be comparable to a previous, more tentative estimate by Uri et al. (Unpublished; 101 seeds.m^{-2}.yr^{-1}). However, they used a lower *Enhalus* density (21 shoots.m^{-2}). Further, considering the results from the previous chapter (Chapter 4) that the actual seedling establishment was only 0.67 seedling.m^{-2}.yr^{-1} , then export and other losses must be considerable (± 99% in total).

DISCUSSION

Spatio-temporal variation in flowering

Temporal factors

The temporal variation in the flowering intensity of *Enhalus acoroides* correlated with water temperature in this study (Figs. 5.1, 5.3). McMillan (1980, 1982) had been successful in inducing (via temperature manipulation, ranging from 20°C to 31°C) 17 other tropical seagrasses but not *Enhalus*. This failure of the species to flower was not only true for his short-term setup (lasting for ca. 5 months) but also for his long-term (1973-1982) cultures. For this, he suggested that probably the temperature tolerance range for flowering of this species is above 31°C. This suggestion does not conform with Bolinao *Enhalus* as it flowered successfully in all field temperature conditions ranging from 26.5°C to 32.5°C (Tables 5.3, 5.4; Figs. 5.1, 5.2, 5.3). For the failure of *Enhalus acoroides* to flower in McMillan's setup, light limitation seems to be a more plausible explanation. The artificial light used for those setups came from cool, fluorescent bulbs with photon flux densities ranging from 70-200 μE.m^{-2}.s^{-1} (McMillan 1980). Even under continuous light situation, this range translates to only 6-17 E.m^{-2}.d^{-1} (on average probably 12 E.m^{-2}.d^{-1}), a light environment that approximates, or might be worse than, the situation at site 1c (see also Fig. 5.5), where *Enhalus* flowering is rare.

The influence of light on the temporal variation in flowering of *Enhalus* may not be evident at sites receiving irradiance levels above the minimum requirement for flowering induction. However, the temporal variation might be important for *Enhalus* growing near the flowering depth (light) limit (Figs. 5.5, 5.7). For instance, shoot #10 (tagged) at S1c was found to have a mature female flower in mid-May 1995 sampling (Table 5.4), coinciding a prior 2-month period (March-April) of total daily PAR (15-day running average) above 15 E.m^{-2}.d^{-1} (Fig. 5.6).

The timing and duration of female flower exposure above water (as influenced by tidal fluctuations and site depth) are probably major factors in the successful pollination and subsequent fruiting. The fact that flowering in general is all year-round and that male flower discharge has no apparent trigger (Fig. 5.9), makes fruiting, to a large extent, dependent on the tidal situation (cf. Fig. 5.4).

Spatial factors
Comparing *Enhalus acoroides* flowering across sites, both sites 1a and 4 had significantly higher (11-12% means) intensities than any of the other sites 5, 6 and 1c (means $\leq$ 5%). These differences can be explained sufficiently by the gradient in light availability at the respective sites (Fig. 5.5). Plants at a clear- water site of ca. 1-meter depth (site 4) have exceptionally higher intensities (see also other related and consistent results, Fig. 5.7) but otherwise, flowering intensity is linearly correlated with PAR availability. Further extrapolation would show that at a mean PAR availability of ca. 12 $E.m^{-2}.d^{-1}$ or lower, *Enhalus acoroides* flowering is no longer possible (Fig. 5.5). The rarity of flowering at site 1c is most probably due to the fact that the mean PAR received at seagrass depth is close to the projected flowering limit (12 $E.m^{-2}.d^{-1}$, Fig. 5.5).

Nutrient limitation has probably only a minor role in causing the spatial differences in *Enhalus acoroides* flowering. On the one hand, the higher flowering at sites 1a and 1b than at site 1c could not be attributed to nutrient factor. At these sites, there were no significant differences in the levels of nutrients both in the water column and in the sediments (Knight 1996; this study, Chapter 2). From Fig. 5.11, it can be concluded that the relatively larger *Enhalus acoroides* plants at site 1c still have enough nutrient resources for, at least, flower production. On the other hand, the fact that *Enhalus* had lower flowering levels at sites 5 and 6 (where nutrient levels may occasionally increase from river run-off and/or wastewaters) contradicts a case of nutrient limitation.

Temperature, the means being similar in all sites (Chapter 2), was also not a factor in explaining spatial differences. Although the minimum and maximum ranges in temperature may differ across sites (i.e., heating at site 1a during low tide vs. no significant change at site 4) and may cause spatial differences in flowering, this study does not suggest so. Also, sites 1a , 5 and 6 were comparable in this respect and yet significant differences in flowering were found.

On the 'occasional-emergence' hypothesis (Troll 1931), this study has shown that *Enhalus* plants at site 1c (2.5-3.0 m datum depth, where 'occasional emergence' never happens) were able, although rarely, to flower (Fig. 5.1; Table 5.4). Also, *Enhalus* plants growing at ca. 1.2 m datum depth (site 4), flowered intensely (Figs. 5.1, 5.2). At this depth, female flowers and/or leaf tips hardly experienced exposure (Fig. 5.4). Fruiting, however, was only common at site 1a.

Flowering frequency: discrepancy between field and reconstructed estimates

While the reconstruction technique is useful when calculating rates and requires considerably less sampling effort, difficulties in interpreting datasets arise in cases when flowers are initiated but, after some time, aborted. This abortion inference was strongly indicated from site 1c samples showing all the clones to have initiated flowering profusely (ca. 4.5 flower scars.yr^{-1}; Table 5.5) and all the 12 standing shoots beared young but fully-developed flowers in between leaf blades. These young flowers rarely mature as can be deduced from the results using random quadrats (Table 5.3; Figs. 5.1, 5.2, 5.3) and from observations that some of the older ones of those young flowers (in site 1c *Enhalus* samples) were already in various stages of decay (brownish, disintegrating easily) at the time of sampling. For other sites, parallel discrepancies can be seen (Table 5.4 vs. Table 5.5). While the frequency of flower scars on the rhizomes of *Enhalus acoroides* is high (ca. 4 flower scars.yr^{-1}), the actual production of mature flowers is considerably less (ca. 1 flower.yr^{-1}). The reconstructed frequency therefore may be treated only as a 'flowering potential' which may be attained occasionally (see tagged shoot # 1, site 5, Table 5.4).

With this limitation, it becomes difficult to extrapolate aspects of fruit and/or seed production based on flower scars. Most likely, this exercise would overestimate these aspects both within a site (site 1a, seed production, see results) and between sites (in view of the result that the high frequency of flower scar on *Enhalus* rhizomes does not vary across sites, Table 5.5, while in fact, flowering does, Table 5.4). A further complication is the fact that distinguishing the occurrence of fruits from rhizome scars is probably impossible. An *Enhalus* shoot may actually produce many flowers but rarely bear fruits (as in the case of site 4; Figs. 5.1, 5.4).

Male flower discharge: a function of tides?

Existing hypotheses on the development and release mechanism of *Enhalus* male flowers do not seem to work within the present study. Troll (1931) disagreed with Svedelius (1904 as cited in Troll 1931) who put forward the primary role of water pressure on the male flower buds release. For Svedelius, a low water situation removes pressure on the flower buds which allows them to rise to the surface. For Troll, pressure reduction may be important not on the flower buds but on the spathe enclosure. He further argued in favor of a tide-pool theory: (1) flowering plants are often found in depressions containing water, though often < 15 cm, throughout low tide, i.e., tide pools; (2) low-tide pools are 10 °C warmer than flood water, and; (3) this temperature rise probably triggers mass flowering. In this situation, water pressure is no longer critical because water level is already very low and the male flowers need not rise to the surface. The opening of the spathe leaves may even be achieved just by active growth (backfolding, etc.). In this present study, male flower discharges were observed in any tidal condition (Fig. 5.9), therefore pressure-reduction hypothesis is doubtful. Also, male flowers were produced and released at sites 1b and 4 where water levels are not shallower than 0.75 m (see Chapter 2 for tidal fluctuations) and where a significant temperature rise of 10 °C (as maybe the case in tide pools) never happens (Fig. 2.3, Chapter 2; Fig. 5.1).

Perhaps, a more critical factor for mass release of male flowers is the production of gas bubbles. Few studies (Troll 1931; Verhoeven 1979; de Cock 1980) reported the production of gas bubbles associated with flowering. Although the function of gas bubbles may vary with different species (in some cases, no function; Van Vierssen et al. 1982), gas bubbles have been observed to help the inflorescence of *Ruppia cirrhosa* to reach the water surface (Verhoeven 1979). If gas bubbles play a similar role for *Enhalus acoroides* (as was also mentioned in Troll 1931), several observations can be explained: (1) no release occurs in the early morning (pers. obs.) because gas storage has been exhausted from dark respiration; (2)

gas saturation inside tissues occurs around noontime till late afternoon, hence, male flower discharge is also most likely to occur around this period (pers. obs.; Brouns & Heijs 1986); and (3) since gas production is primarily light dependent, the effect is about the same everywhere within areas of about comparable PAR availability at depths, hence, release is synchronous (Troll 1931; Brouns & Heijs 1986; pers. obs.)

Annual flowering frequency in *Enhalus acoroides* vs. other species

The observed intensity in the flowering of *Enhalus acoroides* obtained from this study (means for sites ranging from 0.3% to 12% of the shoots present, Table 5.3) is comparable to the values reported for other seagrass species (mean, 9.6%; range, 3.9-17.8%; Gallegos et al. 1992). However, for most species, flowering is a strongly seasonal event, i.e., it only occurs at a certain period of the year, e.g., among others, *Thalassia testudinum, Zostera marina* (Gallegos et al 1992), *Thalassia hemprichii* (Uri et al. Unpublished; pers. obs.), Zostera noltii (Vermaat & Verhagen 1996) and *Cymodocea nodosa* (Caye & Meinesz 1985; Duarte et al. 1994). This was not the case for *Enhalus acoroides* which showed to flower all year-round (Den Hartog 1970; Brouns & Heijs 1986; this study). Hence, on an annual basis, flowering intensity for this species is far higher, e.g., at sites 1a and 1b, an average shoot flowers at least once a year (Table 5.4). Corresponding values obtained from reconstruction techniques (= flowering potential) showed even higher intensities. Uri et al. (Unpublished) reported that of the 42 *Enhalus* shoots sampled, 18 "flowered" (= flower scar, at least initiated) at least once (24 never flowered), and on average, a flowering shoot "flowers" 25% of its lifetime. Calculating further, this means that a flowering shoot flowers (or, attempts to flower) ca. four times a year. Therefore, of the 42 shoots, a total number of 72 "flowering" events were observed, equivalent to an annual flower scar-shoot ratio of 170%. Similar calculations were done for the *Enhalus* clone samples in this study and parallel conclusions were made. The only difference is that, in this study, all the 12 clones and all shoots flowered. Then, extrapolation on an annual basis, would yield a "flowering"-shoot ratio of 400%, a high potential which is sometimes actually attained by some *Enhalus* shoots (tagged shoot # 1, site 5; Table 5.4). Furthermore, also the realized flowering of *Enhalus acoroides* on an annual basis is high relative to other seagrasses (Table 5.4 vis-a-vis Gallegos et al. 1992). This high investment to sexual reproduction does not agree well with the generalized notion that sexual reproduction is a strategy associated more with annual plants while vegetative reproduction is for perennial plants (de Cock 1980; Grace 1993). Relative to other species found in the area, *Enhalus acoroides* is long-lived (i.e., perennial type; Chapter 4; Vermaat et al. 1995), slow-expanding (Chapter 4; Vermaat et al. 1995) and sexually prolific (this study; Uri et al. Unpublished).

LITERATURE CITED

Brouns, J.J.W.M. and H.M.L. Heijs. 1986. Production and biomass of the seagrass *Enhalus acoroides* (L.f.) Royle and its epiphytes. Aquat. Bot. 25: 21-45.

Caye, G. and A. Meinesz. 1985. Observations on the vegetative development, flowering and seedling of *Cymodocea nodosa* (Ucria) Ascherson on the Mediterranean coasts of France. Aquat. Bot. 22: 277-289.

De Cock, A.W.A.M. 1980. Flowering, pollination and fruiting in *Zostera marina* L. Aquat. Bot. 9: 201-220.

Den Hartog, C. 1970. Seagrasses of the world. North Holland Publ., Amsterdam, 275 pp.

Duarte, C.M. and K. Sand-Jensen. 1990. Seagrass colonization: biomass development and shoot demography in *Cymodocea nodosa* patches. Mar. Ecol. Prog. Ser. 67: 97-103.

Duarte, C.M., N. Marbá, N. Agawin, J. Cebrián, S. Enríquez, M. Fortes, M.E. Gallegos, M. Merino, B. Olesen, K. Sand-Jensen, J.Uri and J. Vermaat. 1994. Reconstruction of seagrass dynamics: age determinations and associated tools for the seagrass ecologist. Mar. Ecol. Prog. Ser. 107: 195-209.

Durako, M.J. 1994. Seagrass die-off in Florida Bay (USA): changes in shoot demographic characteristics and population dynamics in *Thalassia testudinum*. Mar. Ecol. Prog. Ser. 110: 59-66.

Durako, M.J. and M.D. Moffler. 1987. Factors affecting the reproductive ecology of *Thalassia testudinum* (Hydrocharitaceae). Aquat. Bot. 27: 79-95.

Estacion, J.S. and M.D. Fortes. 1988. Growth rates and primary production of *Enhalus acoroides* (L.f.) Royle from Lag-it, North Bais Bay, The Philippines. Aquat. Bot. 29: 347-356.

Gallegos, M.E., M. Merino, N. Marba and C.M. Duarte. 1992. Flowering of *Thalassia testudinum* Banks ex König in the Mexican Caribbean: age-dependence and interannual variability. Aquat. Bot. 43: 249-255.

Gallegos, M.E., M. Merino, N. Marba and C.M. Duarte. 1993. Biomass and dynamics of *Thalassia testudinum* in the Mexican Caribbean: elucidating rhizome growth. Mar. Ecol. Prog. Ser. 185: 185-192.

Grace, J.B. 1993. The adaptive significance of clonal reproduction in Angiosperms: an aquatic perspective. Aquat. Bot. 44: 159-180.

Hootsmans, M.J.M. 1994. A growth analysis model for *Potamogeton pectinatus* L. In: W. van Vierssen, M.J.M. Hootsmans and J.E. Vermaat (eds.). Lake Veluwe, a macrophyte-dominated system under eutrophication stress. Geobotany 21. Kluwer, Netherlands, 250-286 pp.

Hootsmans, M.J.M., J.E. Vermaat and W. van Vierssen. 1987. Seed-bank development, germination and early seedling survival of two seagrass species from the Netherlands: *Zostera marina* L. and *Zostera noltii* Hornem. Aquat. Bot. 28: 275-285.

Johnstone, I. 1979. Papua New Guinea seagrasses and aspects of the biology and growth of *Enhalus acoroides* (L.f.) Royle. Aquat. Bot. 7: 197-208.

Knight, D.L.E. 1996. Allocation of nutrients in the seagrass *Enhalus acoroides*. MSc Thesis E.E. 211, International Institute for Infrastructural, Hydraulic and Environmental Engineering, Delft, The Netherlands, 34 pp.

Kruis, F. 1995. Environmental chemistry. Selected analytical methods. EEO95/95/1, International Institute for Infrastructural, Hydraulic and Environmental Engineering, Delft, The Netherlands.

Marba, N., M.E. Gallegos, M. Merino and C.M. Duarte. 1994. Vertical growth of *Thalassia testudinum*: seasonal and interannual variability. Aquat. Bot. 47: 1-12.

McMillan, C. 1980. Flowering under controlled conditions by *Cymodocea serrulata*, *Halophila stipulacea*, *Syringodium isoeftifolium*, *Zostera capensis* and *Thalassia hemprichii* from Kenya. Aquat. Bot. 8: 323-336.

McMillan, C. 1982. Reproductive physiology of tropical seagrasses. Aquat. Bot. 14: 245-258.

Meñez, E., H.P. Calumpong and R.C. Phillips. 1983. Seagrasses from the Philippines. Smithsonian Contrib. Mar. Sci. no. 21, 40 pp.

Mwaiseje, B. 1979. Occurrence of *Enhalus* on the coast of Tanzania, E. Africa. Aquat. Bot. 7: 393.

Pettit, J.M. 1984. Aspects of flowering and pollination in marine organisms. Oceanogr. Mar. Biol. Ann. Rev. 22: 315-342.

Phillips, R.C. and E.G. Meñez. 1988. Seagrasses. Smithsonian Contrib. Mar. Sci. 34.

Terrados, J. 1993. Sexual reproduction and seed banks of *Cymodocea nodosa* (Ucria) Ascherson meadows on the southeast Mediterranean coast of Spain. Aquat. Bot. 46: 293-299.

Troll, W. 1931. Zur morphologie und biologie von *Enhalus acoroides* (Linn. f.) Rich. Flora 125: 427-456.

Uri, J.S., N.S. Agawin, C.M. Duarte, M.D. Fortes, N. Marba and J.E. Vermaat. Unpbublished Manuscript. Flowering in Philippine seagrasses: age-dependence, seasonality and interannual variability.

Van Vierssen, W., R.J. van Wijk and J.R. van der Zee. 1982. On the pollination mechanism of some polyhaline Potamogetonaceae. Aquat. Bot. 14: 339-347.

Verhoeven, J.T.A. 1979. The ecology of *Ruppia*-dominated communities in Western Europe. I. Distribution of *Ruppia* representatives in relation to their autoecology. Aquat. Bot. 6: 197-268.

Vermaat, J.E. and F.C.A. Verhagen. 1996. Seasonal variation in the intertidal seagrass *Zostera noltii* Hornem.: coupling demographic and physiological patterns. Aquat. Bot. 52: 259-281.

Vermaat, J.E., N.S.R. Agawin, C.M. Duarte, M.D. Fortes, N. Marba and J.S. Uri. 1995. Meadow maintenance, growth and productivity of a mixed Philippine seagrass bed. Mar. Ecol. Prog. Ser. 124: 215-225.

Note. In this chapter the words 'flower' and 'inflorescence' have been used as synonyms, although morphologically this is not strictly correct.

Chapter 6

The effects of sediment type, shading and nutrient addition on the survival, morphometrics and biomass of *Enhalus acoroides* (L.f.) Royle seedlings

Abstract. Seeds of *Enhalus acoroides* from a shallow and clear water site were germinated and grown under various *in situ* and *in vitro* experimental conditions. *In situ* (Expt. 1), seedlings differed in settling success and survival rates as well as in shoot heights after 1 year. *In vitro*, two experiments were done: Expt. 2 tested the effect of sediment type, using sediment types found in various sites, and Expt. 3 tests simultaneously the effects of sediment type, shading and nutrient addition. In Expt. 2, the germination of *Enhalus* seeds was high (ca. 98%) and fast (i.e., 'completed' within 5 days after release). Larger seedlings (both in size and biomass) were obtained on mud than on coarse-sand sediments. In Expt. 3, it was shown that all 3 factors had significant main effects on leaf size and biomass and that sediment type and nutrient addition strongly interacted. For root characteristics, the effect of nutrient addition was strongest, significantly reducing root length and biomass regardless of sediment types and shading levels.

INTRODUCTION

Enhalus acoroides (L.f.) Royle probably inhabits the widest range of local environmental conditions compared to other tropical seagrass species occurring in the Indo-West Pacific Region (Den Hartog 1970; Johnstone 1979; Brouns & Heijs 1986; Estacion & Fortes 1988; Erftemeijer 1993; Tomasko et al. 1993). In Bolinao (NW Philippines), this species is able to colonize various habitat types down to ca. 3 m deep (Vermaat et al. 1995; Chapter 2): muddy to coarse-sandy substratum, turbid to clear waters, intertidal pools and splash zones (above zero datum). Across this range in environmental conditions, the corresponding *Enhalus* shoots also differ in size (i.e., total leaf surface area and biomass; Chapter 3) and density (Chapter 4). Often the distribution of *Enhalus* within a site is patchy (Chapter 4; Vermaat et al. 1995; Agawin et al. 1996) and "bare" zones exist adjacent to *Enhalus* meadows (e.g. site 2, Chapter 2). Patchiness might be explained by the fact that this species is quite slow in asexual horizontal expansion (Chapter 4; Vermaat et al. 1995). Probably, this particularly pertains to areas where effective seed rain (i.e., sexual recruitment) is also low, e.g., when the existing population is unable to bear fruits (sites 1c and 4, Chapter 5) or seedlings fail to establish. Yet, at some shallow and clear water sites (S1a, Chapter 5; Agawin et al. 1996) where this species is sexually productive, *Enhalus* distribution is still not homogenous. A potential reason could be that the overall success of the sexual recruitment process (i.e., from seed settling to fully mature plants) is low. From a demographic perspective, there are two major possibilities: (1) dispersal limitation, i.e., the inability of seeds to reach these sites; and (2) poor settling success and subsequent seedling survival, growth and vegetative expansion.

This paper addresses the second possibility in a series of *in situ* and laboratory experiments. Seeds collected from a shallow, clear-water site (1a) where *Enhalus* fruiting is highest (Chapter 5) were germinated and grown in various sites (Chapter 2) with and without *Enhalus* in order to establish regional (between-site) variation in *in situ* seedling survival and growth. This regional variation, if present, could then be causally linked to several naturally covarying environmental factors. Therefore, a series of subsequent laboratory experiments was carried out addressing the effects of sediment type (muddy, sandy, etc.), shading and nutrient addition, being the most likely factors to vary across these sites.

To summarize, the specific objectives of this paper are: (1) to test *in situ* if *Enhalus* seeds from site 1a are able to establish and survive in a number of selected sites, and if they do, to assess differences in seedling characteristics, and (2) to test *in vitro* the effects of important covarying environment variables, in particular sediment type, light (shading) and nutrients on morphometrics and biomass of seedlings (seeds from site 1a as well).

MATERIALS AND METHODS

Experiment 1. Establishment, survival and shoot sizes of *Enhalus acoroides* seedlings grown at various sites

To test (1) if *Enhalus* seeds from site 1a are able to establish themselves and to survive in other field stations, and (2) to assess differences in seedling morphometrics (e.g., mean shoot height) across sites, the following experiment was carried out.

From site 1a, ca. 100 *Enhalus* fruits were collected (July 1994), opened, and all seeds were removed and thoroughly mixed. At each of the sites 1a, 1b, 1c, 2, 3, 5 and 6 (see site descriptions in Chapter 2), one hundred randomly-chosen seeds were germinated and grown in pots. At each site, the seeds were divided into 4 replicate sets of 25 seeds each. Each set was placed in a pot (cylindrical plastic container, ca. 17 cm diameter, 15 cm high) previously filled with intact sediment cores from the same site. During each subsequent sampling event (during July 1994 - October 1995) once every 1-2 months, the number of seedlings and ungerminated seeds were counted. A seed was considered to be germinated when the first foliage leaf had emerged and, from then on, was considered to be a seedling. Mean seedling leaf length was approximated (underwater) by measuring five representative seedlings per pot. At site 2, all seedlings died-off quickly. To test whether the cause of the "quick die-off" was related to the sediment type, 100 more seeds were released (February 1995) in four pots, two of which contained sediments from site 1a while the other two were filled with sediments from site 2.

In most sites, the first 2-month period was considered to be the establishment phase of seedlings, thereafter survival was calculated based on the number of established seedlings. For sites 5 and 6 where sediment top layers were relatively loose (muddy), the period of establishment was extended up to 4 months as seedling losses due to washout at these sites were still high after 2 months. Seedling mortality rates were calculated as the slopes of the lines regressing seedling survival against time (expressed in years). To test for differences between mortality rates, multiple comparisons between regression lines were done by calculating an F-statistic based on the residual sums of squares of separate and pooled lines (Vermaat & Hootsmans 1994; see also Chapter 4).

Experiment 2. *In vitro* test of sediment type effect

To test the effect of the different sediment types on seed germination, seedling morphometrics and biomass, a total of 30 glass tanks (61 x 35 x 46 cm : L x W x H) filled with ca. 10-cm layer of sediments taken from different sites (1a, 1c, 2, 3, 5 and 6) were arranged in an outdoor platform (Fig. 6.1). Five replicate tanks were used for each sediment type (see Chapter 2 for sediment charateristics) collected by taking several sediment cores of ca. 30-cm depth at the sites and subsequently mixing the cores belonging to the same site.

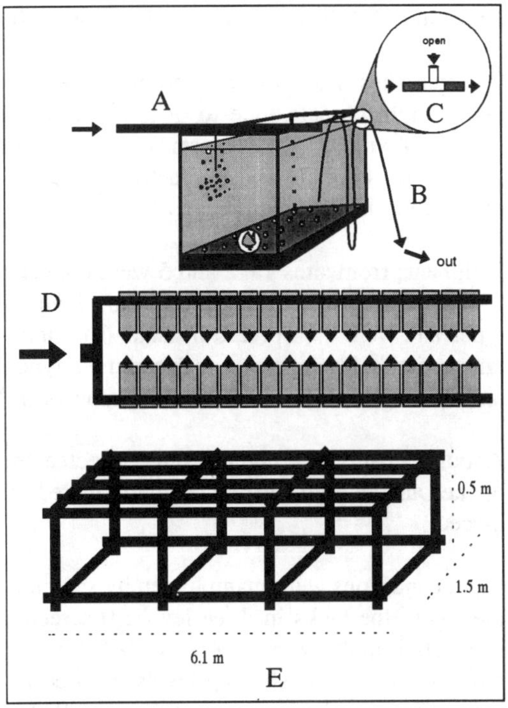

Fig. 6.1. Schematic diagram of the laboratory experimental tank setup. **A** = glass tank, 61 x 35 x 46 cm, L x W x H; **B** = water drainage hose; **C** = T-connector serving as a water level regulator; **D** = tank arrangement with the supply pipes; **E** = outdoor wooden platform.

Seawater from the nearby strait (Fig. 2.1, Chapter 2, Channel near BML; salinity of 33.3 ± 1.8 ‰, mean ± sd; temperature of 28.7 ± 1.25 °C, mean ± sd; Rivera 1997) was supplied into the tanks via the central seawater pumping system of the Bolinao Marine Laboratory of the Marine Science Institute. The supply and drainage system (flow-through) was controlled such that the water in the tanks was replenished twice between 0800H-1800H daily. Seawater supply was off during the night (1800H-0800H).

In December 1994, a total of 15 seeds from three different fruits were released in each tank. Seed planting was done in three clusters to distinguish the fruit-source. The number of germinated seeds was counted every day during the first 15 days (or until the percentage of germinated seeds was 100%). After six months (June 1995), all seedlings were harvested carefully including all roots and leaves intact. Leaf and root morphometrics (leaves: number, width, maximum length, total length; roots: number, maximum length, total length) of seedling samples were measured, after which, samples were oven-dried (60°C for 24 hrs.) and weighed.

The effect of sediment type on the measured parameters was tested using ANOVA treating the seedlings as basic replicates. Further, the nested effects of fruit and tank replications were tested subsequently estimating their contribution to the total variance (Sokal and Rohlf 1981). A Tukey test was applied to do multiple comparisons test among means of the different sediment types. Data were log-transformed when departures from normality and homogeneity of variance were detected.

Experiment 3. Simultaneous effects of sediment type, shading and nutrient addition

Between sites, it is not only sediment composition that considerably varies but also, among others, light climate (due to differences in water depths and turbidity) and nutrient inputs (expectedly more in sites closer to mainland; cf. Chapter 2). The objective of this experiment was to test the main and interactive effects of sediment type, shading and nutrient addition on seedling morphometrics and biomass.

Sediments from sites 1a, 2 and 5 were collected and mixed in the same way as in experiment 2. Pots (cylindrical container: 15 cm, diam.; 7.5 cm, h) were filled with different sediment types and were then placed in tanks (the same outdoor tank setup as in Experiment 2). Each tank contained two pots of each sediment type (hence, six pots per tank). The seawater supply and drainage system used was the same as in Experiment 1.

About 75 fruits of *Enhalus* were collected from site 1a in December 1995. All fruits were opened and all seeds were thoroughly mixed. Into each pot, five randomly selected seeds were placed.

Light conditions were manipulated by shading (using black nylon screens; mesh size = 1 mm) the top of the tanks in three levels: 0 screen layer, 1 screen layer, and 3 screen layers. The sides of all tanks were covered with 1/2-inch styropore boards to eliminate shading effect by adjacent tanks. The shading levels resulted in PAR transmittance values of 78, 54 and 30% respectively, and represented the mean PAR transmittance due to water-column depth and turbidity at sites 1a, 5 and 2. Mean remaining PAR levels were 25 $E.m^{-2}.d^{-1}$, 17 $E.m^{-2}.d^{-1}$ and 10 $E.m^{-2}.d^{-1}$ respectively. For each shading level, eight tanks were used. Slow-release nutrient sticks (8% Total Nitrogen (0.4% organic, 3% nitrate/nitrite, 4.6% ammonium), 8% P_2O_5, 10% K_2O, and 12% organic matter; Agawin et al. 1996) were added to each of the pots in four of the eight tanks per shading level.

In April 1996, all seedlings were harvested. Seedling morphometrics and biomass were determined as in Experiment 2.

Main and interactive effects of sediment, shading and nutrient addition on the morphometrics and biomass of *Enhalus* seedlings were tested using ANOVA. Data were log-transformed when departures from normality and homogeneity of variance were detected (e.g., biomass data).

RESULTS

Establishment, survival and shoot heights of *Enhalus acoroides* seedlings grown at the sites (Expt. 1)

Seedling establishment in the field differed considerably across sites (Fig. 6.2). Germination success was not a reason for these differences as *Enhalus* seedlings were quite able to germinate under all conditions (see also results below; Expt. 2) even when still inside the fruits (pers. obs.). In most sites, seed and/or seedling losses were probably due to water

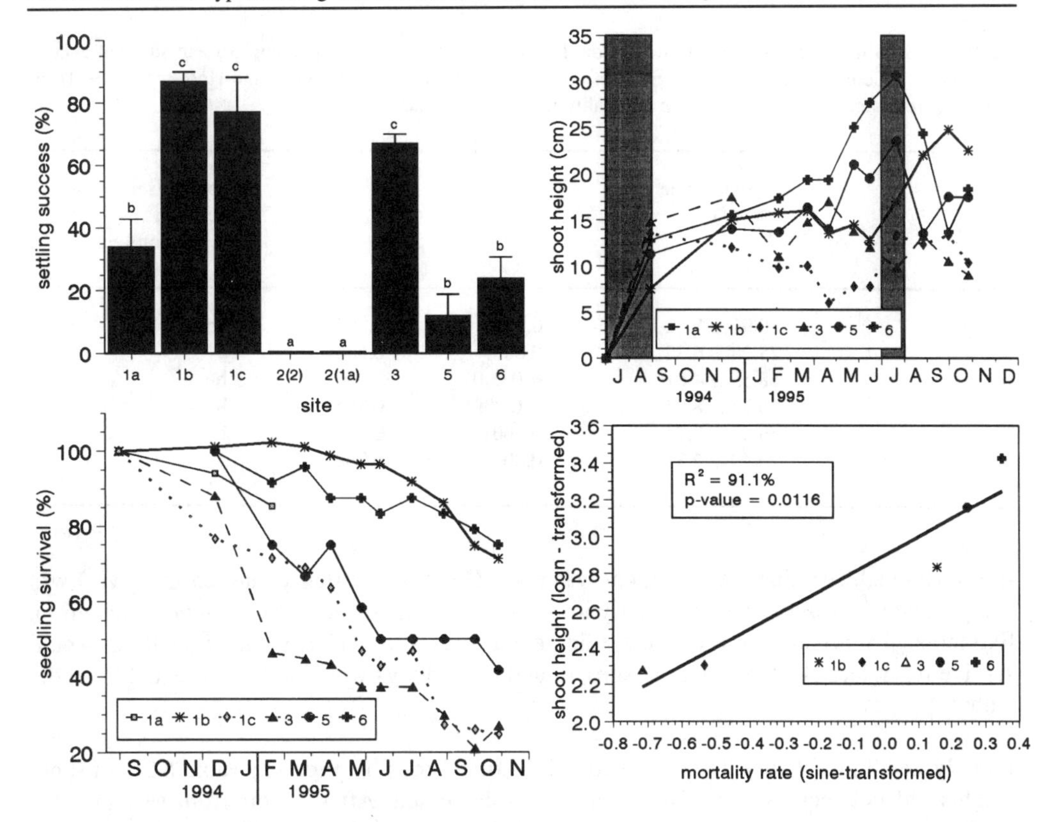

Fig 6.2. Settling success ± se (top) and survival curves (bottom) of *Enhalus* seedlings grown at various sites originating from a shallow, clear-water site (1a). **Top**: on the x-axis are site codes (see Chapter 2); 2(2) = settling success at site 2 using site 2 sediments; 2(1a) = settling success at site 2 using site 1a sediments; different letters above bars indicate sitewise significant differences at $p < 0.05$ (Tukey test); **Bottom**: legends are site codes (see Chapter 2); n = 4 (pots)

Fig. 6.3. Top: Mean ± se shoot heights of *Enhalus* seedlings grown *in situ*. Legends are site codes. The sampling events after 2 months and after 1 year are indicated (shaded bar, see also Table 6.2). Standard errors are not shown for clarity; n = 4 (pots). **Bottom**: mean mortality rates (sine-transformed) plotted against mean shoot heights ($\log_n$ -transformed).

movement (wave energy in sites 1a, 5 and 6; stronger water currents in sites 3 and 1c; pers. obs.) and looseness of the sediment top layer (sites 5 and 6). However, in site 2, establishment was zero neither because seedling losses due to water movement nor looseness of sediment top layer but due to quick mass collapse (death) of seedlings two weeks after seed release. It was initially suspected that sediment characteristics at site 2 might be the reason for the collapse. Therefore the experiment was repeated at this site by also importing sediments from site 1a. The same quick seedling collapse was found (Fig. 6.2).

Seedling survival also differed significantly across sites (Fig. 6.2; Table 6.1). Seedling pots at site 1a were stolen in February 1995 so a survival curve at this site could not be extended any further. At site 1b, seedling survival was best with a mortality rate of 25% yr^{-1} significantly lower than the mortality rates at sites 3, 1c and 5 (59-69% yr^{-1}; Table 6.1). At

Table 6.1. Seedling mortality rates shown as the linear slopes of the lines regressing percent survival against time (in yrs.). Comparisons between regression lines were done following Vermaat and Hootsmans (1994); p-value = partial significance of the slope (mortality rate) of each line.

Site	mortality rate ($\%.yr^{-1}$)	p-value	R^2	comparison between regression lines
1a	30.91 ± 8.32	0.1673	0.865	ab
1b	25.29 ± 6.32	0.0029	0.606	a
1c	68.55 ± 4.61	< 0.0001	0.956	b
3	66.77 ± 8.94	< 0.0001	0.845	b
5	59.44 ± 3.73	0.0001	0.834	b
6	25.49 ± 2.71	0.0001	0.862	a

site 6, although seedling establishment was poor (Fig. 6.2), mortality rate (ca. 25% yr^{-1}) was comparable to 1b and also significantly lower than sites 3, 1c and 5 (59-69% yr^{-1}; Table 6.1). Extrapolated survival curves would predict that at sites 3, 1c and 5, *Enhalus* seedlings would survive for at least 1.7 yrs, while at sites 1b and 6, seedlings would not be expected to all die within 4 years.

Initially, seedlings at sites 3 and 1c were taller (first 2 months; Fig. 6.3; Table 6.2) but shoot heights did not increase any further and even decreased. After a year from seed release, seedlings at site 6 were taller (mean shoot length ca. 30 cm; Fig. 6.3; Table 6.2) than those at the other sites. When plotted against the corresponding mortality rates (sine-transformed; Fig. 6.3, bottom), *Enhalus* seedlings appeared to be taller at the sites where mortality rates were lower.

Table 6.2. Mean (± se) seedling shoot heights of *Enhalus acoroides* at two months and one year after seed release *in situ* (Expt. 1, see also Fig. 6.3). Different letters attached to values indicate sitewise significant differences (per column) at $p < 0.05$ (Tukey test).

	seedling shoot height	
site	after 2 months (cm)	after 1 year (cm)
1a	11.25 ± 1.65 ^{ab}	-
1b	7.5 ± 0.29 ^{a}	17.0 ± 2.68 ^{ab}
1c	13.5 ± 1.7 ^{b}	10.0 ± 3.4 ^{a}
3	14.75 ± 0.75 ^{b}	9.75 ± 2.5 ^{a}
5	11.25 ± 1.18 ^{ab}	23.5 ± 1.5 ^{ab}
6	12.75 ± 1.25 ^{ab}	30.67 ± 6.4 ^{b}

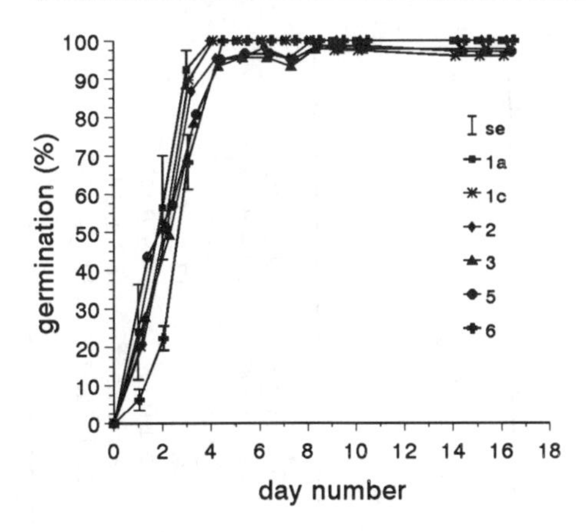

Fig. 6.4. Percent germination of *Enhalus* seedlings at the outdoor laboratory experiment (Expt. 2). For clarity, only the standard errors for sediment types 1a and 6 are shown; n = 5 (tanks).

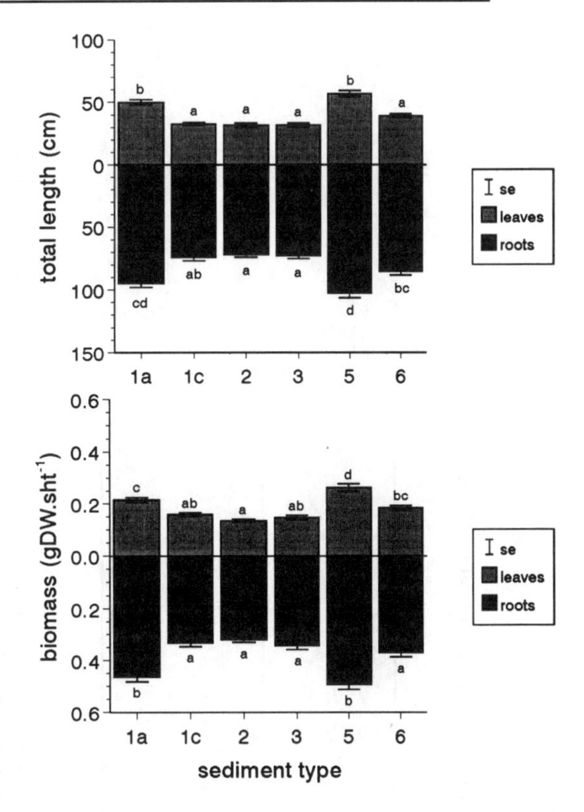

Fig. 6.5. Total length (top) and biomass (bottom) of *Enhalus acoroides* seedlings after 6-month culture (Expt. 2). Different letters attached to bars indicate significant differences across sediment types (Tukey test, $p < 0.05$).

In vitro test of sediment type effect (Expt. 2)

There was no effect of sediment types on seed germination in outdoor aquaria (Fig. 6.4; ANOVA, $p > 0.05$). In all treatments, seed germination was high (98.49 ± 0.87%) and fast (within 5 days after seed release). The effects of sediment type on seedling morphometrics and biomass were significant (Tables 6.3, 6.4). Table 6.3 also shows that the seed and fruit components of the total variance were 37 and 17% respectively, the former was even slightly higher than the sediment type component (28%). Morphometrics (leaves: number, width, length; roots: number, length) and biomass of seedlings grown on site 5 sediment were generally high, while those obtained from sediment types 1c, 2 and 3 were low (Table 6.4; Fig. 6.5). Relative to sediment types 5 and 1a, seedlings grown in sediment type 3 had significantly lower values in most parameters (Table 6.4). Seedlings grown in sediment types 1c and 3 did not differ, in most cases, from those seedlings grown in sediment type 2. However, in some characteristics (e.g., leaf weight and leaf width), seedlings grown in sediment types 1c and 3 were not different from those grown in sediment type 1a. For sediment type 6, seedling characteristics had generally intermediate values.

Simultaneous effects of sediment type, shading and nutrient addition (Expt. 3)

General effects

In all morphometric characteristics (e.g., leaf width, n leaves, length, n roots and root length), all the three main factors (sediment type, shading and nutrient addition) had highly significant effects (see Table 6.5, leaf and root lengths are shown only as representative tests; see also

Table 6.3. Tests of significance (nested ANOVA) and estimation of variance components for total leaf lengths of *Enhalus acoroides* seedlings from Experiment 2.

Source of Variation	SS	DF	MS	F	p	variance component (%)
within cells (among seeds)	48708	354	138			37
between sediment types	47233	5	9447	68.7	< 0.001	28
nested effect: tank	35614	24	1484	10.8	< 0.001	18
nested effect: fruit source	4593	10	459	3.3	< 0.001	17

Table 6.4. Morphometrics of seedlings cultured on different sediment types and harvested after 6 months (Expt. 2). Error values are standard errors. Different letters attached to values (superscript) indicate significant differences ($p < 0.05$; Tukey test) across sediment types. Seeds originated from site 1a. All parameters are expressed per seedling. See also Fig. 6.5 for total length and biomass of below- and above-ground parts.

parameter	sediment type					
	1a	1c	2	3	5	6
leaf width, mm	7.4 ± 0.09^{b}	7.1 ± 0.08^{ab}	6.8 ± 0.10^{a}	7.0 ± 0.10^{ab}	8.1 ± 0.10^{c}	7.5 ± 0.10^{bc}
n leaves	4.5 ± 0.08^{b}	3.7 ± 0.07^{a}	3.8 ± 0.08^{a}	3.7 ± 0.09^{a}	4.4 ± 0.08^{b}	4.0 ± 0.06^{a}
mean leaf L, cm	11.0 ± 0.41^{b}	8.6 ± 0.24^{a}	8.2 ± 0.43^{a}	8.3 ± 0.25^{a}	12.8 ± 0.41^{c}	9.6 ± 0.33^{ab}
max. leaf L, cm	22.3 ± 0.93^{b}	15.0 ± 0.48^{a}	15.5 ± 1.45^{a}	15.3 ± 0.64^{a}	24.7 ± 0.83^{b}	17.3 ± 0.66^{a}
n roots	4.9 ± 0.11^{c}	4.2 ± 0.12^{b}	4.1 ± 0.10^{a}	3.6 ± 0.12^{a}	5.5 ± 0.17^{d}	4.7 ± 0.11^{bc}
mean root L, cm	19.3 ± 0.41^{ab}	17.6 ± 0.51^{a}	18.1 ± 0.58^{a}	20.8 ± 0.51^{b}	19.2 ± 0.56^{ab}	18.3 ± 0.48^{a}
max. root L, cm	30.9 ± 0.63^{c}	27.4 ± 0.75^{ab}	26.5 ± 0.74^{a}	29.8 ± 0.64^{bc}	32.0 ± 0.91^{c}	30.3 ± 0.88^{bc}

Fig. 6.6). Of the three factors, the effect of sediment type on the total leaf length was strongest, explaining ca. 28% of the total variance. For the total root length, the effect of nutrient addition was strongest (34%). All second-order interaction terms (sediment x light; sediment x nutrient; light x nutrient) on root length had significant effects. For leaf length, only the effect of sediment-nutrient interaction was significant. No significant effect of the third-order interaction could be found in all tests.

Similar tests for biomass characteristics yielded parallel results with regards to the main effects (Table 6.5), i.e., the variation in leaf weight was largely (14%) explained by the effect of sediment type while the variation in root weight was largely (27%) explained by the effect of nutrient addition. There was however no main effect of nutrient addition on leaf weight (Table 6.5) but its interaction with sediment type had significant effect. For the rest of the interaction terms, results were in parallel with those obtained for length tests, except that the effect of sediment-light interaction was significant for root weight (Table 6.5). After combining the leaf and root weights into total biomass, only the main effects were significant, each of them explaining an equal proportion of the variation (4-8%).

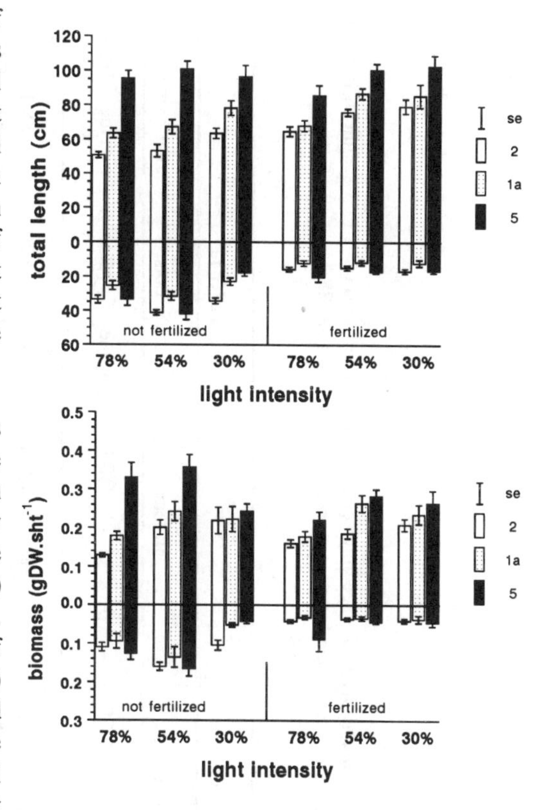

Fig. 6.6. Experiment 3: total length (top) and biomass (bottom) of *Enhalus acoroides* seedlings after 4-month culture; values below 0, belowground parts; values above 0, aboveground parts; legends are sediment type codes; error bars are standard errors. For ANOVA comparisons, see Tables 6.5 and 6.6.

Multiple comparisons

<u>main effects</u>

Consistent with the results from Expt. 2, seedlings grown in site 5 sediments had longer leaf lengths and leaf biomass whilst seedlings in site 2 sediments were the smallest (Fig. 6.6; Table 6.6). Sediment type effect on the roots was less clear, generally no significant effect on root biomass except at 30% light intensity and without fertilization, i.e., longer roots and more root biomass at sediment type 2.

The reduction of light intensity increased leaf length and biomass (Fig. 6.6; Table 6.6) of seedlings grown under sediment types 1a and 2. Seedlings growing in sediment type 5 did not show similar length increases when shaded. Instead, the above-ground biomass was significantly reduced at the highest shading level. The main effect of shading on root length and biomass was significant only when no nutrients were added, in which case, root length

Table 6.5. Three-way ANOVA table for the length and biomass of *Enhalus acoroides* seedlings cultured for 4 months using various sediment, light and nutrient conditions (Expt. 3). %SS = the ratio of the sum of squares (SS) due to the factor (e.g., sediment) against total SS.

Source of variation	LENGTH				WEIGHT					
	all leaves		all roots		all leaves		all roots		total biomass	
	p-value	%SS	p-value	%SS	p-value	%SS	p-value	%SS	p-value	%SS
SEDIMENT	< 0.001	27.69	< 0.001	4.30	< 0.0001	13.51	< 0.0001	2.79	< 0.0001	7.50
LIGHT	< 0.001	4.39	< 0.001	3.67	< 0.0001	3.98	< 0.0001	4.20	< 0.0001	3.86
NUTRIENT	< 0.001	3.09	< 0.001	33.66	0.122	0.48	< 0.0001	27.37	< 0.0001	7.49
SEDIMENT x LIGHT	0.610	0.41	0.001	2.36	0.403	0.80	0.298	0.78	0.135	0.84
SEDIMENT x NUTRIENT	0.001	2.32	0.005	1.34	0.002	2.50	0.048	0.97	0.378	0.84
LIGHT x NUTRIENT	0.092	0.73	< 0.001	4.30	0.857	0.06	0.001	2.24	0.121	0.77
SEDIMENT x LIGHT x NUTRIENT	0.562	0.45	0.057	1.16	0.570	0.58	0.522	0.51	0.718	0.42
basic error (among seedlings)		60.68		49.89		77.96		61.84		81.93

Table 6.6. Multiple comparisons of morphometrics and biomass of *Enhalus acoroides* seedlings cultured for 4 months using various sediment, light and nutrient conditions (Expt. 6.3). Different letters indicate significant differences (Tukey test, $p < 0.05$) across levels of a particular treatment. For example, across sediment types at 78% light intensity and 'not fertilized', leaf lengths were all different from each other, i.e., only comparing group numbers 1, 2 and 3; across nutrient level at 78% light intensity and sediment type 5, group numbers 3 and 12 were compared and the difference was not significant (ns); Different letters also indicate which value is higher or lower, i.e., $a < b < c$.

nutrient addition	not fertilized									fertilized								
light intensity	78%			54%			30%			78%			54%			30%		
sediment type	2	1a	5	2	1a	5	2	1a	5	2	1a	5	2	1a	5	2	1a	5
group number	1	2	3	4	5	6	7	8	9	10	11	12	13	14	15	16	17	18
Morphometrics																		
total leaf length																		
across sediment types	a	b	c	a	b	c	a	a	b	a	a	b	a	b	c	a	ab	b
across light levels	a	a	ns	a	ab	ns	b	b	ns	a	a	ns	b	b	ns	b	b	ns
across nutrient levels	a	ns	ns	a	a	ns	a	ns	ns	b	ns	ns	b	b	ns	b	ns	ns
total root length																		
across sediment types	ns	ns	ns	b	a	b	b	a	a	a	a	b	ab	a	b	ns	ns	ns
across light levels	a	ab	b	b	b	b	a	a	a	ns	ns	ns	ns	ns	ns	ns	ns	ns
across nutrient levels	b	b	b	b	b	b	b	b	ns	a	a	a	a	a	a	a	a	ns
Biomass																		
total leaf biomass																		
across sediment types	a	a	b	a	a	b	ns	ns	ns	a	ab	b	a	b	b	ns	ns	ns
across light levels	a	ns	ab	ab	ns	b	b	ns	a	a	a	ns	ab	b	ns	b	ab	ns
across nutrient levels	a	ns	b	ns	ns	b	ns	ns	ns	b	ns	a	ns	ns	a	ns	ns	ns
total root biomass																		
across sediment types	ns	ns	ns	ns	ns	ns	b	a	a	ns	ns	ns	ns	ns	ns	ns	ns	ns
across light levels	a	ab	b	b	b	b	a	a	a	ns	ns	ns	ns	ns	ns	ns	ns	ns
across nutrient levels	b	b	ns	b	b	b	b	ns	ns	a	a	ns	a	a	a	a	ns	ns

and biomass were reduced at the highest shading level. With fertilization, shading had no effect on root morphometrics and biomass of seedling in all sediment types.

Fertilization positively affected leaf length and biomass of seedlings cultured under sediment types 1a and 2 conditions (Fig. 6.6; Table 6.6). Leaf lengths of seedlings grown under sediment type 5 condition were not sensitive to nutrient addition. Leaf biomass of these seedlings even tended to be lower relative to the unfertilized groups. The effect of nutrient addition on the roots was very clear, i.e., a strong reduction of length as well as biomass regardless of sediment types and shading levels.

interactive effects

As a general trend for leaf length (see Table 6.5 and Fig. 6.6), the positive effect of nutrient addition was dependent on the sediment type, e.g., no effect when plants were grown on sediment type 5. Shading had no interactive effect on leaves, i.e., sediment-light and nutrient-interactions were not significant. For the roots, all the three interaction terms (Tables 6.5, 6.6, Fig. 6.6) were significant: (1) the effect of shading was significant only when no nutrients were added, and its significance was dependent on sediment type; (2) the effect of sediment type was significant in both nutrient conditions but also dependent on both shading and nutrient levels, e.g., at 78% light intensity, no sediment type effect without nutrient addition was found while there was a significant sediment type effect with nutrient addition; at 30% light intensity, there was significant sediment type effect without nutrient addition while none was found when nutrients were added.

DISCUSSION

Settling success and survival

The field experiment on settling success and survival yielded different results at different sites. An immediate mass mortality of seedlings was found at site 2, suggesting that even if seeds are able to reach the site (which is most likely due to its proximity to site 1a, i.e., only 70 m away), seedlings die quickly. The reason for this mass mortality is not known. However, a sediment quality aspect seems unlikely as an explanation since importing sediment from 1a (Expt. 1) also yielded the same seedling collapse. Further tests (Expts. 2 and 3), showed that seedlings grown in site 2 sediment in the laboratory performed less than those grown in other sediment types, but mass mortality was not the case. A water-depth aspect (at site 2, datum depth = 3.5 m) also seems an unlikely explanation considering that seedlings at site 3 (datum depth = 5. m; Expt. 1) survived.

At the unpopulated site 3, mortality in the field was high (~65%.yr^{-1}) but was comparable to 1c (Table 6.1) implying that given enough recruitment (i.e., > 65%.yr^{-1}), *Enhalus* (if established) would probably be able to maintain a population at site 3 as is apparently the case at site 1c.

The gradient in *Enhalus* density decreasing from site 1a to 1c (Rollon et al. Unpublished data), might be caused by (1) a gradient in seed rain as fruits are not produced at 1c (Chapter 5) and (2) a higher mortality rate in the deeper part (Table 6.1), or both.

Variation in morphometric characteristics and biomass across sites: role of sediment type, light and nutrients

Results from the field (Expt. 1) and laboratory (Expts. 2 and 3) experiments showed that seedling lengths and biomass were: (1) higher at site 5 sediment relative to 1a, (2) enhanced by shading in sediment types 1a and 2, and (3) increased by nutrient addition. Root lengths and biomass however significantly decreased in all sediment and shading combinations.

These results fit well with what can be observed from the field. The larger *Enhalus* size and biomass at site 5 compared with site 1a are apparent effects of sediment type as was consistently demonstrated in Expts. 2 and 3. At site 1c where sediment type itself is expected to produce smaller plants (Expt. 2), still, *Enhalus* plants are larger than at the shallower 1a and 1b sites of the site 1 transect (Chapter 3). This can be explained by the low-light environment prevailing at the site (1c) in view of its deeper water-column. Shading (e.g., water-column) promotes larger plants grown from seeds as exhibited by Expt. 3. At site 6, plants are also larger relative to 1a (Chapter 3). At this site sediment characteristics are not much different from 1a (Chapter 2) but the water-column is more turbid (so, lower-light) and, probably, water-column nutrients are periodically enriched, by virtue of the fact that site 6 was located in front of the town's fish landing and market site, and adjacent to the main town center (Chapter 2, Fig. 2.1).

The considerable variation (biomass and shoot size) shown by seedlings originating from one site within one experiment is rather comparable to the variation among the natural populations present at sites 1a, 1b, 1c, 5 and 6. This strongly suggests that the size variation observed in *Enhalus* populations in Bolinao may be largely a phenotypic response to a varying environment.

The remarkable decrease in root length and biomass in any sediment type when nutrients were added (Fig. 6.4) indicates the active role played by roots in the acquisition of nutrients. Energy allocation to roots (or below ground parts) in seagrass has also been demonstrated elsewhere to correlate inversely with nutrient status (Perez et al. 1991; Hemminga et al. 1994; Agawin et al. 1996).

Seagrass communities (i.e., mature plants) have been documented to be negatively affected by shading (Vermaat et al. 1993; Philippart 1995; Gordon et al. 1994; Abal et al. 1994; Van Lent et al. 1995) which results contrast with the findings in the present study on seedlings. In general, seedlings are known to actively respond to light reduction through a mechanism called etiolation (Went & Sheps 1969; Vermaat & Hootsmans 1991). However, the *in situ* shading of mature plants significantly reduced growth and biomass of *Enhalus acoroides* (Chapter 7). Apparently, seedlings responses differ from that of adult plants, probably due to (a) the carbon stored in the endosperm, and (b) low respiratory needs.

Till to date, two contrasting papers report on nutrient limitation on SE Asian seagrasses: one showed a limited effect (Erftemeijer et al. 1994) while the other found a strong evidence (Agawin et al. 1996). The latter (Agawin et al. 1996) reconciled this disagreement as largely methodological. Results from Expt. 3 showing a strong interaction between sediment type and nutrient addition suggest that both conclusions (Erftemeijer et al. 1994 vs. Agawin et al. 1996) are possible from doing *in situ* nutrient-addition experiments depending on the sediment characteristics at the site. At sites with sediment types comparable to that of sites 5 and 6 (i.e., terrigenous), nutrient addition will probably have no effect while stronger effects can be expected at sites with sediments comparable to that of site 1a (i.e., carbonaceous, for example, Agawin et al. 1996; 3 other sites in this study). On the reef flat around Santiago Island (see Chapter 2: Fig. 2.1), seagrasses are found mostly at sites with sediment characteristics comparable to that of site 1a. So, nutrient-limitation may be the general rule here. However, in a wide area further south, sediments resemble those of site 5 (see also Terrados et al. in prep.) and the seagrass beds, often dominated by *Enhalus*, are probably less nutrient-limited. As a general extrapolation, one might postulate that Philippine seagrass beds on terrigenous, silty sediments are less likely to be nutrient-limited than those on carbonaceous beds of autochtonous marine origin.

LITERATURE CITED

Abal, E.G., N. Loneragan, P. Bowen, C.J. Perry, J.W. Udy and W.C. Dennison. 1994. Physiological and morphological responses of the seagrass *Zostera capricornii* Aschers. to light intensity. J. Exp. Mar. Biol. Ecol. 178: 113-129.

Agawin, N.S.R., C.M. Duarte and M.D. Fortes. 1996. Nutrient limitation of Philippines seagrasses (Cape Bolinao, NW Philippines): *in situ* experimental evidence. Mar. Ecol. Prog. Ser. 138: 233-243.

Brouns, J.J.W.M and H.M.L. Heijs. 1986. Production and biomass of *Enhalus acoroides* (L.f.) Royle and its epiphytes. Aquat. Bot. 25: 21-45.

Den Hartog, C. 1970. Seagrasses of the world. North Holland, Amsterdam, 275 pp.

Erftemeijer, P.L.A. 1993. Factors limiting growth and production of tropical seagrasses: nutrient dynamics in Indonesian seagrass beds. PhD Dissertation, Katholieke Universiteit Nijmegen, The Netherlands, 173 pp.

Erftemeijer, P.L.A., J. Stapel, M.J.E. Smekens and W.M.E. Drossaert. 1994. The limited effect of *in situ* phosphorus and nitrogen additions to seagrass beds on carbonate and terrigenous sediments in South Sulawesi, Indonesia. J. Exp. Mar. Biol. Ecol. 182: 123-140.

Estacion, J.S. and M.D. Fortes. 1988. Growth rates and primary production of *Enhalus acoroides* (L.f.) Royle from Lag-it, North Bais Bay, The Philippines. Aquat. Bot. 29: 347-356.

Gordon, D.M., K.A. Grey, S.C. Chase and C.J. Simpson. 1994. Changes to the structure and productivity of a *Posidonia sinuosa* meadow during and after imposed shading. Aquat. Bot. 47: 265-275.

Hemminga, M.A., B.P. Koutstaal, J. van Soelen and A.G.A. Merks. 1994. The nitrogen supply to intertidal eelgrass (*Zostera marina*). Mar. Biol. 118: 223-227.

Johnstone, I. 1979. Papua New Guinea seagrasses and aspects of the biology and growth of *Enhalus acoroides* (L.f.) Royle. Aquat. Bot. 7: 197-208.

Perez, M., J. Romero, C.M. Duarte and K. Sand-Jensen. 1991. Phosphorus limitation of *Cymodocea nodosa* growth. Mar. Biol. 109: 129-133.

Philippart, C.J.M. 1995. Effects of shading on growth, biomass and population maintenance of the intertidal seagrass *Zostera noltii* Hornem. in the Dutch Wadden Sea. J. Exp. Mar. Biol. Ecol. 188: 199-213.

Rivera, P.C. 1997. Hydrodynamics, sediment transport and light extinction off Cape Bolinao, Philippines. PhD Dissertation, IHE-WAU, The Netherlands, 244 pp.

Sokal, R.R. and F.J. Rohlf. 1981. Biometry. W.H. Freeman and Company, New York. 859 pp.

Terrados, J., C.M. Duarte, M.D. Fortes, J. Borum, N.S.R. Agawin, S. Bach, U. Thampanya, L. Kamp-Nielsen, W.J. Kenworthy, O. Geertz-Hansen and J.E. Vermaat. In prep. Changes in community structure and biomass of seagrass communities along gradients of siltation in SE Asia. To be submitted to Mar. Ecol. Prog. Ser.

Tomasko, D.A., C.J. Dawes, M.D. Fortes, D.B. Largo and M.N.R. Alava. 1993. Observations on a multi-species seagrass meadow off-shore of Negros Occidental, Republic of the Philippines. Bot. Mar. 36: 303-311.

Van Lent, F., J.M. Verschuure and M.L.J. van Veghel. 1995. Comparative study on populations of *Zostera marina* L. (eelgrass): *in situ* nitrogen enrichment and light manipulation. J. Exp. Mar. Biol. Ecol. 185: 56-76.

Vermaat, J.E. and M.J.M Hootsmans. 1991. Growth of *Potamogeton pectinatus* L. in a temperature-light gradient. In: M.J.M. Hootsmans and J.E. Vermaat. Macrophytes, a key to understand changes caused by eutrophication in shallow freshwater systems. PhD Dissertation, IHE-WAU, The Netherlands, 27-55 pp.

Vermaat, J.E. and M.J.M. Hootsmans. 1994. Intraspecific variation in *Potamogeton pectinatus* L.: a controlled laboratory experiment. In: W. van Vierssen, M.J.M. Hootsmans and J.E. Vermaat (eds.). Lake Veluwe, a macrophyte-dominated system under eutrophication stress. Geobotany 21, Kluwer Academic Publishers, Dordrecht, The Netherlands, 26-39 pp.

Vermaat, J.E., J.A.J. Beijer, R. Gijlstra, M.J.M. Hootsmans, C.J.M. Philippart, N.W. van den Brink and W. van Vierssen. 1993. Leaf dynamics and standing stocks of intertidal *Zostera noltii* Hornem. and *Cymodocea nodosa* (Ucria) Ascherson on the Banc d'Arguin (Mauritania). Hydrobiologia 258: 59-72.

Vermaat, J.E. N.S.R. Agawin, C.M. Duarte, M.D. Fortes, N. Marba and J.S. Uri. 1995. Meadow maintenance, growth and productivity of a mixed Philippine seagrass bed. Mar. Ecol. Prog. Ser. 124: 215-225.

Went, F.W. and L.O. Sheps. 1969. Environmental factors in regulation of growth and development: ecological factors, In: F.C. Steward (ed). Plant physiology- a treatise, Vol. 5a. Academic Press, NY, 229-406 pp.

Chapter 7

On the survival, morphology and growth of mature stands of the seagrass *Enhalus acoroides* (L.f.) Royle: *in situ* imposed shading and transplantation experiments

Abstract. Two months of imposed 70% shading significantly affected shoot size and growth rate of *Enhalus acoroides* and reduced leaf biomass by ca. 30%. Other size parameters were also reduced in various magnitudes ranging from 10-50%. Absolute growth rate was reduced by ca. 40%. Shoot density was unaffected by shading for four months.

Transplanting mature *Enhalus* individuals from a shallow- and clear-water site to darker (deep or turbid waters) sites yielded comparable establishment success (ca. 40%) with the exception of one site where immediate mass mortality occurred. The transplants to the shallow, clear-water site from other *Enhalus* populations were more successful in establishment (> 70%). Of the parameters measured, leaf width was the most plastic, i.e., the differences in leaf widths of plants transplanted from different populations to one site were lost (converged), while the leaf widths of plants transplanted to other sites from one population (shallow, clear-water) became different (diverged).

INTRODUCTION

The preceding chapters (Chapters 3, 4, 5 and 6) demonstrated the significant influence of light on growth, plant size, and reproduction of *Enhalus acoroides* (L.f.) Royle. Light not only affected the morphology and growth of the natural stands of *Enhalus acoroides* but is also critical in its sexual reproductive allocation. The plants in darker environments (deep and clear as well as shallow but turbid waters) were larger (Chapter 3) but were less abundant (Chapter 4) and produced less flowers (Chapter 5) than those found at a clear-shallow water site. In fact, in deep water (3 m) vegetation, flowering is a rare event which is most likely due to light limitation (Chapter 5). It has also been shown that the temporal variation in *Enhalus* leaf growth can be predicted reasonably by a photosynthesis model based on PAR availability (Chapter 3).

The results (Chapter 3) showing contrasting effects of light availability on the spatial and temporal variation in *Enhalus* size need a logical explanation. A plausible hypothesis would be that the variation in size across sites (e.g., sites with different water depths but with otherwise similar environmental characteristics) is a result of light acclimation during plant development (i.e., larger plant size as an adaptation to low light condition; Went & Sheps 1969, Vermaat & Hootsmans 1991) while, once fully mature, the seasonal variation in size and growth positively correlates with the temporal variation in light. More specifically, it appears as if low light experienced during an early stage of the development into a mature plant ontogenetically orients the plants towards the attainment of a probably adaptive larger size. Conversely, it may also be that in these harsh environments, seedlings which survive are only those that have the largest available pool of reserve material, a selection mechanism for large seeds.

The results of the *in vitro* experiment testing the effect of shading on *Enhalus* seedling characteristics (Chapter 6) support the corollary hypothesis on seedlings. Seedlings grown under low light were larger than those grown under high light condition. For the mature *Enhalus*, no evidence obtained from direct experimental manipulations, so far, exists. For some other seagrass species, the reduction in abundance and biomass of seagrass stands as a consequence of artificial shading has been demonstrated (Backman & Barilotti 1976; Bulthuis 1983; Neverauskas 1988; Vermaat et al. 1993; Abal et al. 1994; Gordon et al. 1994; Fitzpatrick & Kirkman 1995; Philippart 1995). However, different seagrass species respond to light reduction over different time scales, i.e., some species may respond to a short-term shading while others may show sensitivity over long-term tests (e.g., Abal et al. 1994).

The present study aims to determine the morphological and growth responses of mature *Enhalus* stands to *in situ* light reduction which was, herein, manipulated in two ways: (1) artificially shading mature stands in a shallow, clear-water (depth ca. 1 m, K_d = 0.3 m^{-1}) meadow for a relatively short-term period (1-4 months), and (2) transplanting mature plants from a shallow and clear-water site to darker (deep, clear; shallow, turbid, K_d ca. 0.6- 0.7 m^{-1}) environments. Transplantation sites included the two deep sites (3 and 5 m datum depths, K_d = 0.3 m^{-1}) where no *Enhalus* naturally occurs (Chapter 2). This experiment also aimed to test whether *Enhalus* plants would be able to survive at these sites. If so, then, the absence of *Enhalus* may be causally attributed to recruitment limitation.

Further, morphological and growth responses of mature *Enhalus acoroides* to an improved light climate were tested. To do so, mature plants from darker environments were transplanted to a shallow and clear-water site. For the *Enhalus* transplants from a population where flowering is a rare event, special attention was given to the production of sexual reproductive structures.

MATERIALS AND METHODS

Imposed shading

In the period 05 July 1994 - 31 August 1994, two plots with dense *Enhalus* populations were selected at a shallow, clear-water site (datum depth = 1 m; Kd = 0.3 m^{-1}; comparable with site 1b, Chapter 2): one plot was shaded with 3 layers of black polyamide screens (mesh size: 1 mm) reducing PAR by 70% (see Table 7.1 for absolute PAR remaining); the other plot was used as a control. The shading structure resembled an inverted box (2 m x 2 m x 1.5 m : L x W x H), also reducing light coming in from the sides. The shading structure was kept vertically upright by attaching floats on top. In the period 26 January 1995 - 11 April 1995, the experiment was repeated, and, another shading design - a raft type (allowing free water flow through the

Table 7.1. Mean (± sd) PAR at the *in situ* shading setups. Means were obtained by averaging the daily summations of PAR (see Chapters 2 and 3) during the experimental periods.

period	unshaded ($E.m^{-2}.d^{-1}$)	70% shaded ($E.m^{-2}.d^{-1}$)
05 Jul 1994 - 31 August 1994	14.87 ± 4.46	4.46 ± 1.97
26 Jan 1995 - 11 April 1995	24.04 ± 7.21	6.17 ± 1.85

bed which might be a problem in the box type), was installed as well. The raft type consisted of three layers of shading material (as in the box type) fixed to a rectangular frame (3 m x 3 m). This assembly was then fixed at midwater depth (just above the *Enhalus* leaf tips) by ropes and pegs. Floats were attached to keep the structure hovering over the vegetation.

In all cases, 20 *Enhalus* plants (replicates) from different plots were randomly selected and marked with thin wires. Holes were punched through all the leaves. After 10-15 days, these marked plants were harvested and leaf lengths, leaf widths, growth rate and biomass per shoot were measured at the laboratory, following the method described in Chapter 3. Samplings were done before shading, i.e., day 0, and subsequently at days 10, 20, 40 and 60 during the experimental period.

In the analysis, comparisons (ANOVA, Sokal & Rohlf 1981) were made between mean values obtained from: (1) shaded and controls; (2) different shading types, and; (3) box-type results of July-August 1994 and January-April 1995 experiments. For clarity in graphical presentations, only the differences between the shaded and the control values (expressed as percentages of the corresponding control values) are shown.

In the period November 1995 to March 1996, the effect of 70% shading on the shoot density of *Enhalus acoroides* was tested. This was done by selecting six plots (3 x 3 m each plot) at a shallow, clear-water site (datum depth = 1 m; Kd = 0.3 m^{-1}) site. Three of these plots were shaded by a raft-type structure (as above). The other three plots were used as controls. In each plot, 9 units of 0.25 m^2 quadrats (metal frames) were tightly fixed. The total number of *Enhalus* shoots in each quadrat was counted at the start (November 1995) and end (March 1996) of the experiment. The effect of shading on the change in shoot density (start vs. end) was tested (ANOVA).

Transplantation

In February 1994, cores (0.5 x 0.5 m by ca 0.4 m sediment depth) of *Enhalus* plants were collected from shallow, clear-water site 1a (see site descriptions in Chapter 2). Each core may consist of several *Enhalus* clusters, each cluster may consist of several connected shoots (Tomlinson 1974; Duarte et al. 1994; Vermaat et al. 1995a) ranging from 1-6 (mean ca. 3) shoots per cluster (Chapter 4). Several cores were made to have at least a total of 20 *Enhalus* shoots ready for transplantation to each of the sites 1c, 2, 3, 5 and 6 (see site descriptions in Chapter 2). *Enhalus* is not naturally present at sites 2 and 3. Controls were also made by translocating ca. 20 plants within site 1a. For site 2 where a quick mass collapse of transplants occurred, 60 more plants were similarly collected from 1a and transplanted to site 2. The same quick collapse was observed.

Within the same week, at each of the sites 1c , 5 and 6, at least a total of 20 plants was collected the same way as above, and transplanted to site 1a. Corresponding controls (transplants within sites) were also made.

During each sampling (once every 1-2 months), the surviving number of shoots were counted. Establishment success was expressed as the ratio (after 1 month) of the number of surviving shoots with respect to the number of shoots transplanted. The rate of decrease onwards was considered as the mortality rate. Aside from counting the surviving number of shoots, growth rates and morphometrics (leaf width, no. of leaves, leaf lengths) were measured *in situ*. Growth parameters were measured and expressed as in Chapter 3. New vegetative shoots and flowers produced were also recorded.

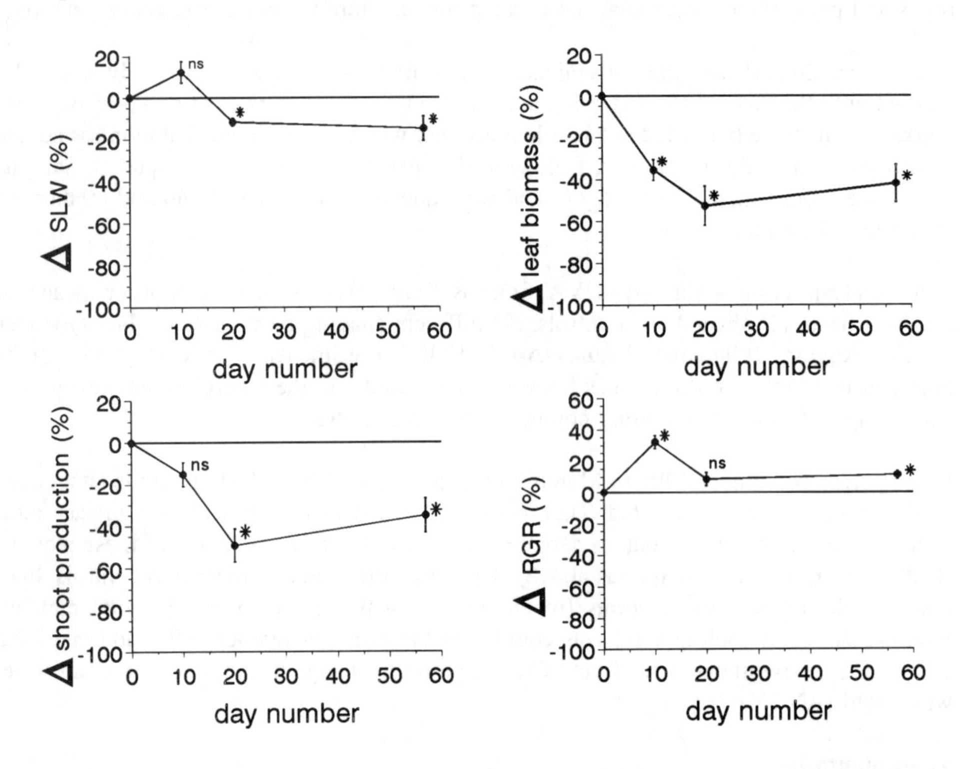

Fig. 7.1. The effect of 70% shading in July-August 1994 on the specific leaf weight (SLW, $mgDW.cm^{-2}$), biomass ($mgDW.sht^{-1}$), shoot production ($mgDW.sht^{-1}.d^{-1}$) and relative growth rate (RGR, $\%.sht^{-1}.d^{-1}$) of *Enhalus acoroides*. The curves depict differences (percentages) against the control, e.g., values lower than 0 show reductions of the parameter by shading. Asterisks (*) indicate significant differences (ANOVA, $p < 0.05$) when comparing the absolute values (i.e., shaded vs. control); ns = not significant.

RESULTS

Imposed shading

Short-term 70% shading reduced the aboveground biomass and production of *Enhalus acoroides* by ca. 40% (Figs. 7.1 and 7.2). In general, the magnitude of the shading effect was less pronounced in the July-August 1994 experiment than in the January-April 1995 experiment (Fig. 7.1 vs. Fig. 7.2), especially when comparing the specific leaf weight (SLW, ca. 10-15% reduction vs. 40-55% reduction, respectively) and the relative growth rate (RGR, positive vs. negative effect). These differences in magnitude could be attributed to the greater absolute amount of PAR reduction during the January-April experiment (18 vs. 10 $E.m^{-2}.d^{-1}$; see Table 7.1). In both cases (Figs. 7.1 and 7.2), the effect of 70% shading was immediate

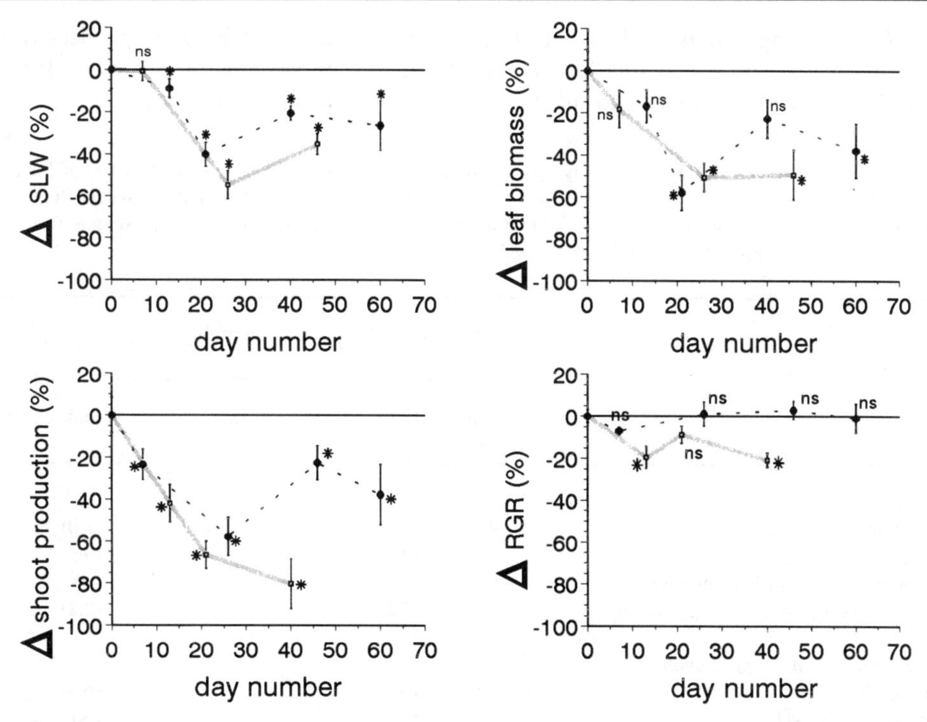

Fig. 7.2. The effect of 70% shading in January-April 1995 on the specific leaf weight (SLW, $mgDW.cm^{-2}$), biomass ($mg.sht^{-1}$), shoot production ($mgDW.sht^{-1}.d^{-1}$) and relative growth rate (RGR, $\%.sht^{-1}.d^{-1}$) of *Enhalus acoroides*. See Fig. 7.1 for curve and legend explanations. The dashed curves represent the difference between the controls and the 'raft-type' shading design; the solid curves = control vs. box-type.

Table 7.2. Significance test (ANOVA) of the change in *Enhalus acoroides* density after *in situ* shading (70%) for 4 months (November 1995 - March 1996). No effect by shading and/or blocking was detected.

Source of Variation	DF	Mean Square	F	Signif of F
Main Effects	3	10.701	0.862	0.468
SHADING	1	6.328	0.510	0.479
BLOCK	2	12.689	1.022	0.368
SHADING x BLOCK	2	4.233	0.341	0.713
Explained	5	8.113	0.654	0.660
Residual	44	12.413		
Total	49	11.974		

(i.e., within 10 days after shading) and no strong indication was present that prolonged shading would magnify the effect. Both 'box-' and 'raft-type' shading designs yielded significantly lower values than the corresponding controls with the exception of RGR parameter (Fig. 7.2).

While 70% shading had immediate effects on shoot size and leaf growth, the density of *Enhalus acoroides* remained unaffected after 4 months of shading (Table 7.2): density remained 7.6 ± 0.4 shoots per 0.25 m^2 (mean ± se; n = 54 quadrats, pooled) .

Table 7.3. Establishment success after 30 days (1 month) and subsequent mortality rate (1.5 years) of *Enhalus acoroides* transplants within sites as well as among sites. Establishment and mortality indices are ratios relative to transplants within 1a; ∞ = cannot be determined; the high mortality value (309%) for transplants from site 1a to site 5 means that after ca. 4 months all individuals were lost (probably washed out).

transplants	establishment success		mortality rate	
	value (%)	(index)	value ($\%.yr^{-1}$)	(index)
within sites				
1a - shallow (0.5 m), clear	44	(1.0)	9 ± 4	(1.0)
1c - deep (3 m), clear	0	(0.0)	∞	(∞)
5 - shallow (0.5 m), turbid, muddy	0	(0.0)	∞	(∞)
6 - shallow (0.5m), turbid, sandy-muddy	91	(2.1)	20 ± 6	(2.2)
from high (1a) to low light climate				
1c - deep (3 m), clear	25	(0.6)	58 ± 31	(6.4)
2 - deep (3.5 m), clear	0	(0.0)	∞	(∞)
3 - deep (5 m), clear	70	(1.6)	49 ± 5	(5.4)
5 - shallow (0.5 m), turbid, muddy	37	(0.8)	309 ± 19	(34.3)
6 - shallow (0.5 m), turbid, sandy-muddy	45	(1.0)	4 ± 16	(0.4)
from low to high (1a) light climate				
1c - deep (3 m), clear	100	(2.3)	23 ± 7	(2.6)
5 - shallow (0.5 m), turbid, muddy	70	(1.6)	58 ± 15	(6.4)
6 - shallow (0.5 m), turbid, sandy-muddy	79	(1.8)	19 ± 5	(2.1)

Transplantation

Enhalus transplants to site 2 did not establish. The first batch of transplants (25 shoots) all died within 2 weeks after transplantation. The additional 60 transplants suffered the same fate. Transplants to site 3 established comparatively well (70% success, Table 7.3). The subsequent rate of decrease (49% yr^{-1}, Table 7.3) of the surviving shoots was comparable with those of other transplants (e.g., 1a-1c, 5-1a, Table 7.3) but leaf width, leaf length and growth rate were clearly deteriorating (Figs. 7.3 and 7.4). Transplants to site 3 (and also to 1c) were unable to produce new vegetative shoots, contrasting especially with those transplants within, from and to site 6.

Comparing the establishment success of the within-site transplants (controls), transplants within site 6 were most successful (91%). Relative to the within-site transplants at 1a, the establishment success of transplants within site 6 is two-fold better (Table 7.3). Plants translocated within sites 1c and 5 did not establish. In the case of 1c, the transplants died-off

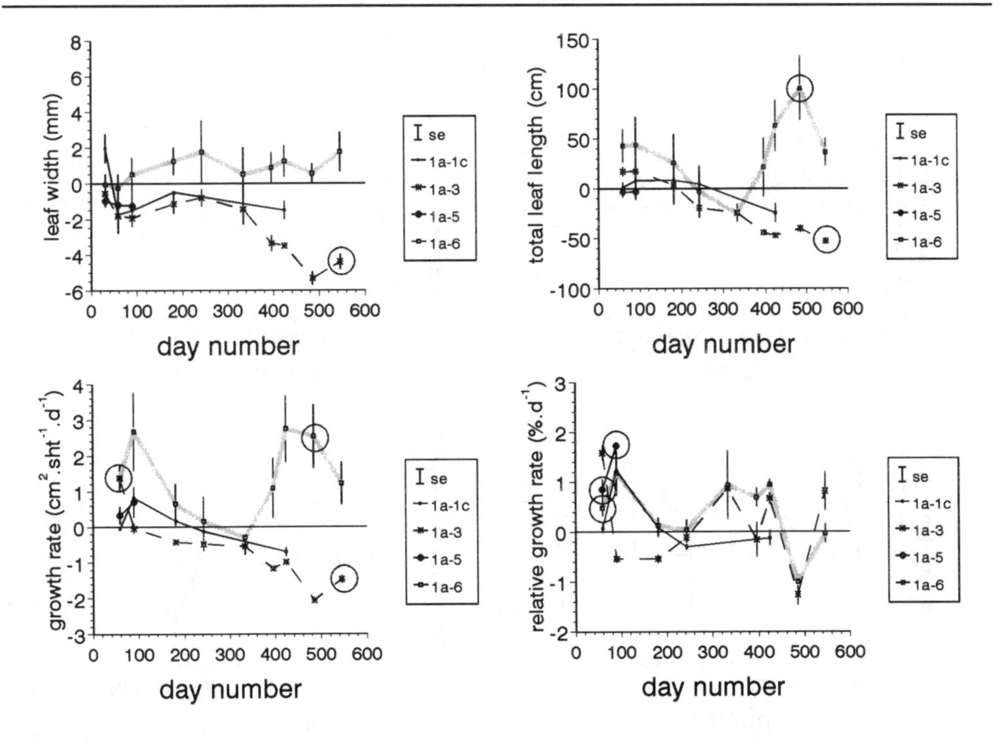

Fig. 7.3. Differences in leaf width (mm), total leaf length (cm), absolute growth rate ($cm^2.sht^{-1}.d^{-1}$) and relative growth rate (RGR, $\%.sht^{-1}.d^{-1}$) of the surviving *Enhalus acoroides* transplants from site 1a to other sites compared to transplants within 1a. Circles indicate significant differences against the control tested at the start (after ca. 60 days) and end (after ca 500 days) of the transplantation experiment. See also Table 7.4.

within 2 weeks after transplantation. At site 5, transplants were washed-out due probably to the unconsolidated nature of the sediment (loose, silty top layer).

Relative to transplants within site 1a, transplants to 1c had a lower success (25%, success index ca. 0.6; Table 7.3) while those transplanted to sites 5 and 6 were similar (ca. 40%, index ca. 1; Table 7.3). Transplants to 1a from sites 1c and 5 (where within site transplants did not establish) had high success (100% and 70% respectively; Table 7.3). Transplants from site 6 to 1a relative to within site 6, showed comparable success.

To generalize: *Enhalus acoroides* transplanted from a shallow and clear-water site (1a) to darker environments (sites 1c, 2, 3, 5 and 6) were less successful in establishing (success index $\leq$ 1, mean = 0.8; Table 7.3) than vice versa (success index $\geq$ 1.5, mean = 1.9; Table 7.3).

Of the surviving *Enhalus* transplants from a shallow, clear-water site (1a) to sites 1c, 3, 5 and 6, plants at deepest (datum depth = 5 m; K_d ca. 0.3 m^{-1}) site 3 showed the clearest change (Fig. 7.3; Tables 7.4 and 7.5). *Enhalus* leaves became much narrower, shorter and produced less amount of leaf parts (AGR) than the controls at 1a. In contrast, successfully established transplants from site 1a to a shallow but turbid site 6 (datum depth ca. 0.5 m; K_d ca. 0.7 m^{-1})

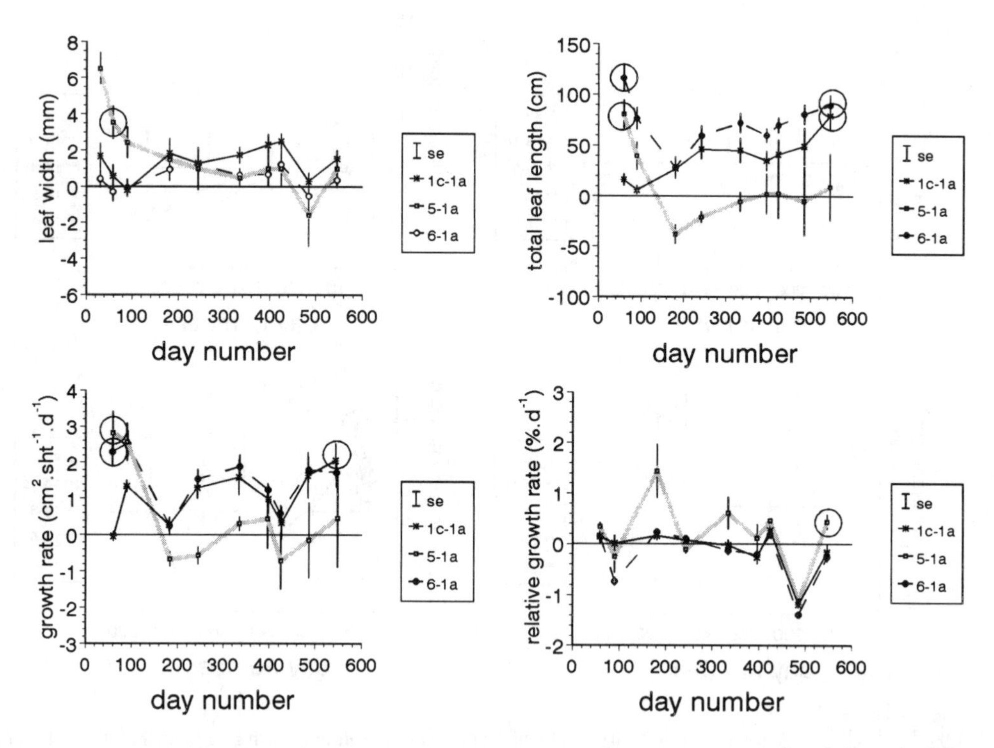

Fig. 7.4. Differences in leaf width (mm), total leaf length (cm), absolute growth rate ($cm^2.sht^{-1}.d^{-1}$) and relative growth rate (RGR, $\%.sht^{-1}.d^{-1}$) of the surviving *Enhalus acoroides* transplants to site 1a from other sites compared to transplants within 1a. For other details, see Fig. 7.3.

kept more and taller leaves resulting in greater total leaf length and higher leaf production (Fig. 7.3).

Of the transplants to site 1a from each of the sites 1c, 5 and 6, *Enhalus* from sites 1c and 5 exhibited. the clearest response, showing immediate decreases in leaf widths and total leaf lengths (Fig. 7.4; Tables 7.4 and 7.5). For *Enhalus* from site 1c (datum depth = 3 m; Kd = 0.3 m^{-1}) which normally have wider and taller leaves (Chapters 2 and 3), decreases in leaf width and total leaf length occurred within a month after transplantation. After ca. 546 days, plants became again taller than the controls (transplants within site) at site 1a. A more gradual but most remarkable leaf width change was exhibited by *Enhalus* transplants from a shallow (datum depth ca. 0.5 m) and more turbid (K_d ca. 0.6 m^{-1}) site 5 to site 1a (Fig. 7.4; Tables 7.4 and 7.5). From an average width of ca. 20 mm at the time of transplantation, leaf width decreased continuously and, after ca. 4 mos., a significant difference from 1a could no longer be detected.

A complete reciprocal comparison can be made between 1a and 6 since both within-site transplants (controls) survived. In terms of leaf width, total leaf length and absolute growth rate, transplants to and from site 6 were performing better relative to the transplants within 1a, while apparently worse relative to transplants within site 6 (Fig. 7.5; Table 7.3).

Table 7.4. *Enhalus acoroides* transplant characteristics compared at days 60 (ca. 2 months) and 546 (ca. 1.5 yrs.; final sampling) after transplantation (see also Figs. 7.3, 7.4 and 7.5). Transplants from 1a to sites 1c and 5 which all died before final sampling, values before death were compared with the corresponding 1a-1a values (in parenthesis). Asterisks (*) attached to values indicate significant difference at $p < 0.05$ (ANOVA) from which it is compared, e.g., 1a-1a vs 5-1a leaf width means that 13.75 ± 0.63 mm < 17.25 ± 0.89 mm; ns = not significant. Transplant ID uses site coding, for instance, 1a-1c refers to transplants from site 1a to site 1c while 5-1a refers to transplants from site 5 to site 1a; AGR = leaf production rate; RGR = relative growth rate.

transplant ID	leaf width (mm)		total leaf length ($cm.sht^{-1}$)		AGR ($cm^2.sht^{-1}.d^{-1}$)		RGR ($\%.sht^{-1}.d^{-1}$)	
	day 60	day 546	day 60	day 546	day 60	day 546	day 60	day 546
one-way transplants								
1a-1a vs.	13.75 ± 0.63	12.00 ± 1.04	45.00 ± 9.50	69.17 ± 11.57	0.99 ± 0.24	1.84 ± 0.44	1.6 ± 0.18	2.15 ± 0.05
1a-1c	12.00 ± 1.00 ^{ns}	9.50 ± 0.50 ^{ns}	46.50 ± 4.50 ^{ns}	33.75 ± 9.25 ^{ns}	0.94 ± 0.35 ^{ns}	0.49 ± 0.12 ^{ns}	1.7 ± 0.09 ^{ns}	1.56 ± 0.15 ^{ns}
		(11.00 ± 2.00)		(58.00 ± 20.00)		(1.19 ± 0.65)		(1.69 ± 0.13)
1a-3	11.93 ± 0.47 ^{ns}	7.6 ± 0.40 *	62.07 ± 4.33 ^{ns}	16.44 ± 2.82 *	2.37 ± 0.18 *	0.37 ± 0.10 *	3.22 ± 0.14 *	2.96 ± 0.38 ^{ns}
1a-5	12.92 ± 0.66 ^{ns}	11.75 ± 0.77 ^{ns}	41.54 ± 4.18 ^{ns}	41.19 ± 7.32 ^{ns}	1.33 ± 0.23 ^{ns}	2.19 ± 0.41 ^{ns}	2.47 ± 1.00 *	4.36 ± 0.06 *
		(13.00 ± 2.89)		(44.33 ± 9.58)		(1.48 ± 0.65)		(2.62 ± 0.77)
1c-1a	14.35 ± 0.62 ^{ns}	13.55 ± 0.47 ^{ns}	60.54 ± 5.51 ^{ns}	123.51 ± 12.41 *	1.45 ± 0.12 ^{ns}	3.54 ± 0.47 *	1.78 ± 0.13 ^{ns}	1.99 ± 0.13 ^{ns}
5-1a	17.29 ± 0.89 *	13.00 ± 1.00 ^{ns}	124.86 ± 13.75 *	53.25 ± 32.75 ^{ns}	4.29 ± 0.60 *	1.94 ± 1.30 ^{ns}	1.96 ± 0.08 ^{ns}	2.58 ± 0.14 *
reciprocal transplants								
1a-1a vs.								
1a-6	13.50 ± 0.77 ^{ns}	13.75 ± 1.09 ^{ns}	87.80 ± 16.80 ^{ns}	105.43 ± 13.37 ^{ns}	2.38 ± 0.33 *	3.06 ± 0.59 ^{ns}	2.11 ± 0.15 ^{ns}	2.11 ± 0.19 ^{ns}
6-1a	13.47 ± 0.52 ^{ns}	12.37 ± 0.40 ^{ns}	160.70 ± 12.85 *	133.70 ± 9.87 *	3.78 ± 0.37 *	3.21 ± 0.39 ^{ns}	1.77 ± 1.00 ^{ns}	1.88 ± 0.08 ^{ns}
6-6 vs.	11.36 ± 0.41	14.08 ± 0.83	141.55 ± 7.80	172.25 ± 12.36	2.94 ± 0.36	5.45 ± 0.49	1.87 ± 0.10	2.26 ± 0.07
1a-6	13.50 ± 0.77 *	13.75 ± 1.09 ^{ns}	87.80 ± 16.80 ^{ns}	105.43 ± 13.37 *	2.38 ± 0.33 ^{ns}	3.06 ± 0.59 *	2.11 ± 0.15 ^{ns}	2.11 ± 0.19 ^{ns}
6-1a	13.47 ± 0.52 *	12.37 ± 0.47 *	160.70 ± 12.85 ^{ns}	133.70 ± 9.87 *	3.78 ± 0.37 ^{ns}	3.21 ± 0.39 *	1.77 ± 1.00 ^{ns}	1.88 ± 0.08 *

Table 7.5. Morphological and growth indices of *Enhalus acoroides* transplants relative to transplants within site 1a (groups 1 & 2) and within site 6 (group 3). Each index is a ratio between the corresponding value (e.g. leaf width) and the value obtained from within site (see also Table 7.4). For example, leaf width index of 0.87 at day 60 obtained for transplants 1a-1c calculated as 12.00 ÷ 13.75 (see Table 7.4), and thus, indicates that transplants to 1c became thinner. Asterisks indicate that the difference between the absolute values (see also Table 7.4) is significant at $p < 0.05$ (ANOVA).

transplant ID	leaf width		total leaf length		AGR		RGR	
	day 60	day 546	day 60	day 546	day 60	day 546	day 60	day 546
1. from lighter (1a) to darker environments								
1a-1c	0.87	0.79	1.03	0.49	0.95	0.27	1.06	0.73
1a-3	0.82	0.63*	1.38	0.24*	2.37*	0.20*	2.01*	1.38
1a-5	0.94	0.98	0.92	0.60	1.34	1.19	1.54*	2.03*
1a-6	0.98	1.03	1.95	1.52	2.40*	1.66	1.32	0.98
mean	0.90	0.86	1.32	0.71	1.76	0.83	1.48	1.28
2. from darker evironments to 1a								
1c-1a	1.04	1.13	1.35	1.79*	1.46	1.92*	1.11	0.93
5-1a	1.26*	1.08	2.77*	0.77	4.33*	1.05	1.23	1.20*
6-1a	0.98	1.03	3.57*	1.93*	3.82*	1.74	1.11	0.87
mean	1.09	1.08	2.56	1.50	3.20	1.57	1.15	1.00
3. indices relative to 6-6 transplants								
1a-6	1.19*	0.98	0.62	0.61*	0.81	0.56*	1.13	0.93
6-1a	1.19*	0.87*	1.14	0.78*	1.29	0.59*	0.95	0.83*

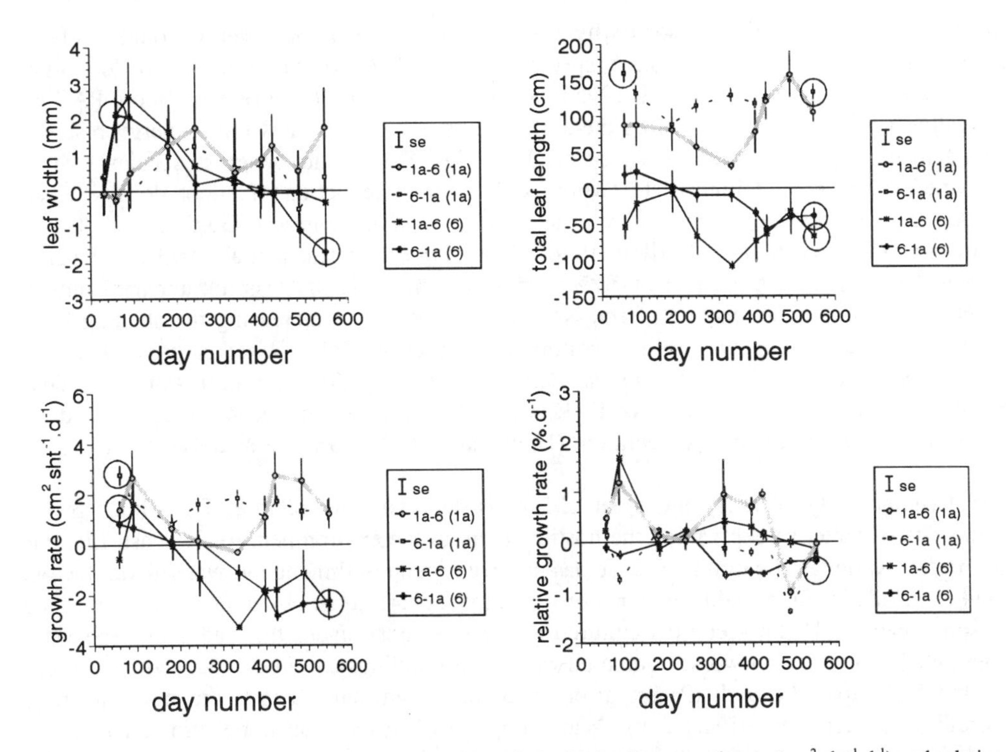

Fig. 7.5. Differences in leaf width (mm), total leaf length (cm), absolute growth rate ($cm^2.sht^{-1}.d^{-1}$) and relative growth rate (RGR, $\%.sht^{-1}.d^{-1}$) of the surviving *Enhalus acoroides* transplants from site 1a to site 6 and vice versa. Legend explanation, e.g.,: 1a-6 (6) = transplants from 1a to 6 compared against transplants within site 6; 1a-6 (1a) = transplants from 1a to 6 compared against 1a. For circles, see Fig. 7.3.

To summarize, *Enhalus* transplants from lighter to darker environments had, on average, lower morphological and growth indices than vice versa (Table 7.5) suggesting that the latter were doing better.

Transplants within and from 1a to other sites did not produce flower during the entire study period. Of the transplants to 1a, plants from sites 1c and 6 simultaneously produced flowers in July 1995 (1c: 1 ♂, 2 ♀ and 6: 2 ♀).

DISCUSSION

Light reduction by 70% shading in this study appeared to have immediate effects on the shoot size, biomass and growth of *Enhalus acoroides*. These effects were already apparent within days of shading. This result supports the finding that light variation (which might be due to cloud cover, tides, water column, turbidity and periphyton) has a major impact on the seasonality in growth of the mature stands of this species (Chapter 3). However, shoot density remained unaffected after 4 months of shaded condition. This result is remarkable compared with other coexisting species such as *Syringodium isoetifolium*, *Cymodocea rotundata* and *Halodule uninervis*. Shoots of these species were wiped-out within a month of 70% shading (pers. obs.). The probable reasons why the long-lived *Enhalus acoroides* would hold-out longer than short-lived, small other species are: (1) *Enhalus* has a relatively low light compensation point (Vermaat et al. 1995b; Agawin et al. 1996); (2) it receives more PAR by extending its longer leaves towards the surface, and (3) *Enhalus* has more energy reserves stored in its larger rhizomes (Vermaat et al. 1995a). Some other seagrass species studied elsewhere also showed immediate (within 0-4 months) reduction of shoot density when shaded, e.g., *Zostera capricorni* (Abal et al. 1994), *Posidonia sinuosa* (Gordon et al. 1994), *Zostera marina* (Backman & Barilotti 1976), *Zostera noltii* (Vermaat et al. 1993). In the case of *Posidonia sinuosa*, the effect of 50% shading on shoot density was not apparent until 6 months (i.e., prolonged shading effect, Neverauskas 1988) but increasing the shading level to 80-90% affected density within 3 months (Gordon et al. 1994). Whether prolonged (i.e., > 4 months) shading will reduce *Enhalus* density can not be inferred from this present study. Furthermore, the absolute amount of PAR (which is not always given in the above studies) is the relevant quantity for consideration (Hootsmans et al. 1993; see also Table 7.1).

The mass mortality of transplants at site 2 is consistent with the collapse of seedlings grown at this site (Chapter 6) confirming the hostile nature of the environment for, at least, *Enhalus acoroides*. Although the ultimate cause has not been found, sediment and water depth factors can be excluded from possible explanations. Sediment characteristics can be eliminated as the reason because: (1) the seedlings cultured at the laboratory under this sediment type were doing well although a bit worse compared with those seedlings cultured under sediment types 1a and 5 (Chapter 6), and; (2) importing sediments from site 1a to 2 did not stop mass mortality of seedlings (Chapter 6). Water depth (3.5 m) can also be eliminated as an explanation, considering that both seedlings (Chapter 6) and transplants at a deeper, nearby site (5 m, site 3) survived for at least a year although were clearly deteriorating (Figs. 7.3; 6.3; 6.4).

The result that transplants survived for at least a year at the deepest site (5 m, site 3) supports the earlier hypothesis (Chapter 6) that given enough seedling recruitment rate, *Enhalus acoroides* could maintain a population at this site. It has been estimated that an annual recruiment rate of > 65% should maintain an *Enhalus* population at the site (Chapter 6).

The transplants within 1c did not establish. This may suggest that at this site (which was only 10 m away from site 2), there exists a delicate balance within the *Enhalus* root-shoot system and that little disturbance can cause death. This might partly be the reason why the 10 "permanent" *Enhalus* plants marked at site 1c for long-term growth and flowering measurements died-off (Chapter 5). The most probable cause of that localized death was the holes (pers. obs.) made by Thalassinid shrimps (*Callianasa* sp.) underneath the sediment where plants were growing. However, uprooting *Enhalus* plants from site 1c and transplanting them to a more favorable site (i.e., 1a) yielded the highest establishment (Table 7.1). Furthermore, three of the surviving transplants (1c to 1a) produced healthy and mature ♀ and ♂ flowers in July 1995. *Enhalus* plants at site 1c hardly produce flowers (Chapter 5).

Enhalus transplants showed wide plasticity in leaf characteristics exhibiting changes much sooner than e.g., *Thalassia testudinum* (Van Tussenbroek 1996). Transplants from 1a to other sites, diverged in leaf width, total leaf length and leaf production rate (Fig. 7.3). Transplants from other sites to 1a showed the reverse trend but the extents of changes towards the mean values at the new environment (site 1a) showed to be also dependent on the source (population) of the transplants. Transplants from site 5 appeared to approach completely the values at 1a (Fig. 7.4). Transplants from site 6 to 1a became smaller than their original size but remained larger than the transplants within site 1a (Fig. 7.5). Vice versa, transplants from 1a to 6 became larger than their original size but remained smaller than those transplants within site 6 (Fig. 7.5). These results show a relatively wide plasticity of *Enhalus acoroides* induced by environmental change, supporting the finding of chapter 6. Apparently, a period of 1.5 years is enough for full leaf width adjustment.

LITERATURE CITED

Abal, E.G., N. Loneragan, P. Bowen, C.J. Perry, J.W. Udy and W.C. Dennison. 1994. Physiological and morphological responses of the seagrass *Zostera capricornii* Aschers. to light intensity. J. Exp. Mar. Biol. Ecol. 178: 113-129.

Backman, T.W. and D.C. Barilotti. 1976. Irradiance reduction: effects on standing crops of the eelgrass *Zostera marina* in a coastal lagoon. Mar. Biol. 34: 33-40.

Bulthuis, D.A. 1983. Effects of *in situ* light reduction on density and growth of the seagrass *Heterozostera tasmanica* (Martens ex. Aschers.) den Hartog in Western Port, Victoria, Australia. J. Exp. Mar. Biol. Ecol. 67: 91-103.

Cambridge, M.L. and A.J. McComb. 1984. The loss of seagrass in Cockburn Sound, Western Australia. 1. The time course and magnitude of seagrass decline in relation to industrial development. Aquat. Bot. 20: 229-243.

Cambridge, M.L., A.W. Chiffings, C. Brittan, E. Moore and A.J. McComb. 1986. The loss of seagrass in Cockburn Sound, Western Australia. 2. Possible cause of seagrass decline. Aquat. Bot. 24: 269-285.

Duarte, C.M., N. Marba, N. Agawin, J. Cebrian, S. Enriquez, M.D. Fortes, M.E. Gallegos, M. Merino, B. Olesen, K. Sand-Jensen, J. Uri and J. Vermaat. 1994. Reconstruction of seagrass dynamics: age determinations and associated tools for the seagrass ecologist. Mar. Ecol. Prog. Ser. 107: 195-209.

Fitzpatrick, J. and H. Kirkman. 1995. Effects of prolonged shading stress on growth and survival of seagrass *Posidonia australis* in Jervis Bay, New South Wales, Australia. Mar. Ecol. Prog. Ser. 127: 279-289.

Gordon, D.M., K.A. Grey, S.C. Chase and C.J. Simpson. 1994. Changes to the structure and productivity of a *Posidonia sinuosa* meadow during and after imposed shading. Aquat. Bot. 47: 265-275.

Hootsmans, M.J.M., J.E. Vermaat and J.A.J. Beijer. 1993. Periphyton density and shading in relation to tidal depth and fiddler crab activity in intertidal seagrass beds of the Banc d'Arguin (Mauritania). Hydrobiologia 258: 73-80.

Neverauskas, V. 1988. Response of a *Posidonia* community to prolonged reduction in light. Aquat. Bot. 31: 361-366.

Philippart, C.J.M. 1995. Effects of shading on growth, biomass and population maintenance of the intertidal seagrass *Zostera noltii* Hornem. in the Dutch Wadden Sea. J. Exp. Mar. Biol. Ecol. 188: 199-213.

Sokal, R. and F.J. Rohlf. 1981. Biometry. The principles and practice of statistics in biological research, 2nd ed. WH Freeman and Co., New York, 869 pp.

Tomlinson, P.B. 1974. Vegetative morphology and meristem dependence - the foundation of productivity in seagrasses. Aquaculture 4: 107-130.

Van Tussenbroek, B.I. 1996. Leaf dimensions of transplants of *Thalassia testudinum* in a Mexican Caribbean reef lagoon. Aquat. Bot. 55: 133-138.

Vermaat, J.E. and M.J.M Hootsmans. 1991. Growth of *Potamogeton pectinatus* L. in a temperature-light gradient. In: M.J.M. Hootsmans and J.E. Vermaat. Macrophytes, a key to understand changes caused by eutrophication in shallow freshwater systems, PhD Dissertation, WAU-IHE, The Netherlands, 27-55 pp.

Vermaat, J.E., J.A.J. Beijer, R. Gijlstra, M.J.M. Hootsmans, C.J.M. Philippart, N.W. van den Brink and W. van Vierssen. 1993. Leaf dynamics and standing stocks of intertidal *Zostera noltii* Hornem. and *Cymodocea nodosa* (Ucria) Ascherson on the Banc d'Arguin (Mauritania). Hydrobiologia. 258: 59-72.

Vermaat, J.E., C.M. Duarte and M.D. Fortes. 1995b. Latitudinal variation in life history patterns and survival mechanisms in selected seagrass species, as a basis for EIA in coastal marine ecosystems (Final Report). Project EC DG XII-G CI1*-CT91-0952, IHE Delft, The Netherlands, 38 pp. + 2 annexes.

Vermaat, J.E. N.S.R. Agawin, C.M. Duarte, M.D. Fortes, N. Marba and J.S. Uri. 1995a. Meadow maintenance, growth and productivity of a mixed Philippine seagrass bed. Mar. Ecol. Prog. Ser. 124: 215-225.

Went, F.W. and L.O. Sheps. 1969. Environmental factors in regulation of growth and development: ecological factors. In: F.C. Steward (ed). Plant physiology- a treatise, Vol. 5a. Academic Press, NY, 229-406 pp.

Chapter 8

General discussion and conclusions: the primary importance of light availability on the biology of *Enhalus acoroides* (L.f.) Royle

INTRODUCTION

The main theme of this dissertation is to gain insights into the general question of "how do seagrass meadows respond to environmental changes?". Focusing on the most common and structurally dominant seagrass species in the Indo-West Pacific region, *Enhalus acoroides* (cf. Chapter 1), each of the preceeding chapters (2-7) contributed to elucidate the three main research lines of this dissertation: (1) describing growth characteristics of *Enhalus* in the established phase; (2) linking such characteristics in relation to prevailing environmental conditions; and (3) determining the response of propagules to environmental changes.

This chapter integrates the main findings of the preceeding ones and will attempt to add the 'predictive' research line (3). In the next section, I will argue that variation in light availability alone is sufficient to explain most spatio-temporal variation in morphology, leaf growth, shoot abundance and floral biology of *Enhalus acoroides*. This chapter then proceeds to speculate upon the impact of environmental changes which, for the Philippine coastal waters, will most likely be siltation and/or eutrophication (EMB Report 1990; Rivera 1997). In that section, I will speculate that although the modification of sediment composition and the increase of nutrient levels may both have direct positive effects on the performance of the *Enhalus acoroides*, the negative effect of light climate deterioration (as a consequence of increased water turbidity and periphyton cover) would be more substantial.

As final sections of this chapter, a number of relevant future research lines are presented, afterwhich the major conclusions of this dissertation are listed.

THE PRIMARY IMPORTANCE OF LIGHT AVAILABILITY FOR MORPHOLOGY, LEAF GROWTH AND FLORAL BIOLOGY OF *ENHALUS ACOROIDES*

The greater percentage of the spatio-temporal variation in size, growth and reproduction of *Enhalus acoroides* can be explained largely by differences in light availability. It has been demonstrated that the larger shoot size in darker environment is a general response of developing plants to light reduction (Chapter 3; Chapter 6; see also Went & Sheps 1969, Vermaat & Hootsmans 1994, Ericsson 1995, Short et al. 1995, Middelboe & Markager 1997). But while the differences in light availability across sites have direct proportionality with shoot sizes, the mean relative leaf growth rate (RGR, which is the amount of leaves produced per unit shoot size) remains comparable across sites (2.34 ± 0.11 $\%.d^{-1}$, Table 3.2, Chapter 3; see also Short et al. 1995). This spatial constancy in RGR has several implications to other aspects in the biology of *Enhalus acoroides* inhabiting different environments. Calculations of net photosynthesis as a function of PAR availability at seagrass depths (Chapter 3)

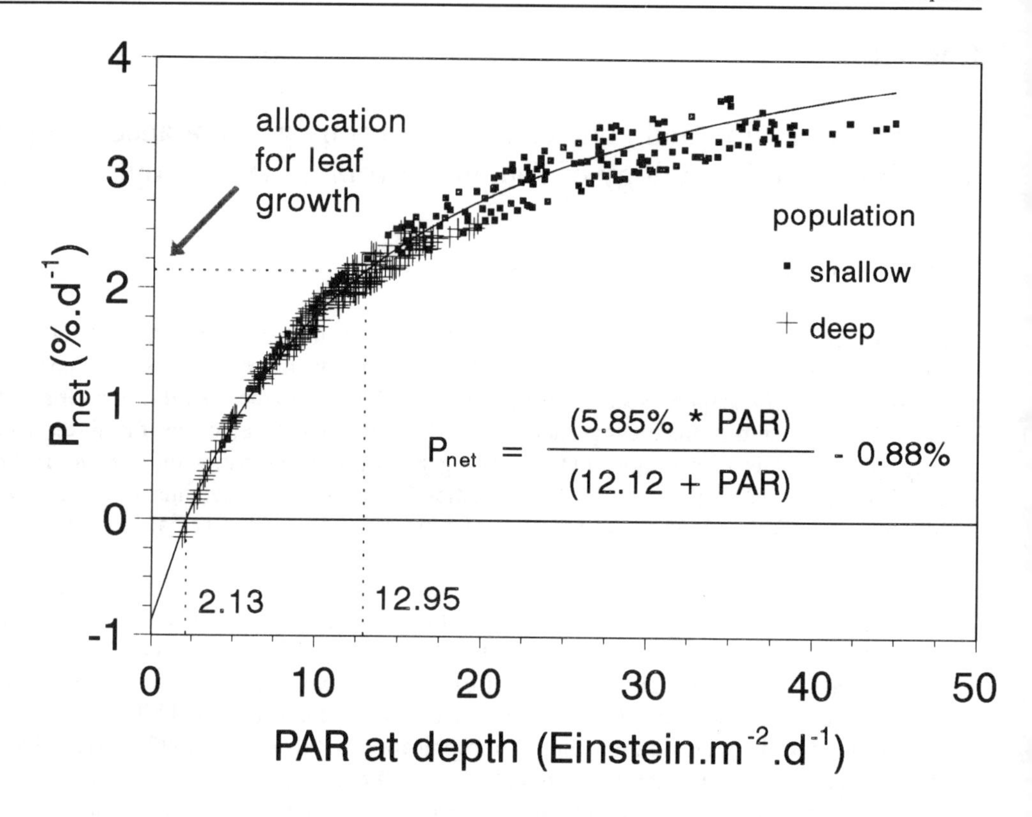

Fig. 8.1. Daily summation of net photosynthesis (P_{net}) plotted against the daily summation of PAR available for *Enhalus acoroides* at depth. The smooth curve defined by the given equation was derived from daily P_{net} and PAR summations at shallow (1a) and deep (1c) populations for the period 01 January 1993 to 31 December 1995 (see Chapter 3 for calculation details). For clarity, the graph shows data points only for the period 01 July 1994 - 31 December 1994. The mean allocation for leaf growth (mean RGR for sites 1a, 1b and 1c ca. 2.15%.d^{-1}, Table 3.2, Chapter 3) and the corresponding PAR requirement (12.95 E.m^{-1}.d^{-1}) are indicated. This PAR level also appears to be the threshold value required for flowering (see also Fig. 5.5, Chapter 5). The PAR level at which P_{net} = 0 (i.e., LCP = 2.13 E.m^{-2}.d^{-1}) is also indicated.

predicted that net photosynthesis could range from -0.25 %.d^{-1} to 2.5 %.d^{-1} in the darkest environment (deep site 1c, mean ca. 2%.d^{-1}) while in the clearest environment (shallow site 1a) , the range is broader and occurs at a much higher level (0.75 %.d^{-1} to 3.75 %.d^{-1}; mean = 3.25 %.d^{-1}). It is remarkable that in the darkest environment (site 1c), the mean RGR and the theoretical net photosynthesis are about equal (Fig. 8.1; see also Fig. 3.5, Chapter 3), implying (1) a priority of leaf growth over other aspects of the repertoire, and, as a consequence, (2) a limitation in carbon and/or energy allocation to sink activities such as the production of flowers, fruits/seeds and/or below-ground biomass (see Fig. 8.2; see also Gifford & Evans 1981, Marschner 1986, Ericsson 1995, Ingestad & Ågren 1995). The saturating form of the curve linking P_{net} to PAR (Fig. 8.1) further implies that the correlation between net photosynthesis and RGR is stronger in environments with low light than in environments with high light levels because (1) the curve is more sensitive at the lower PAR

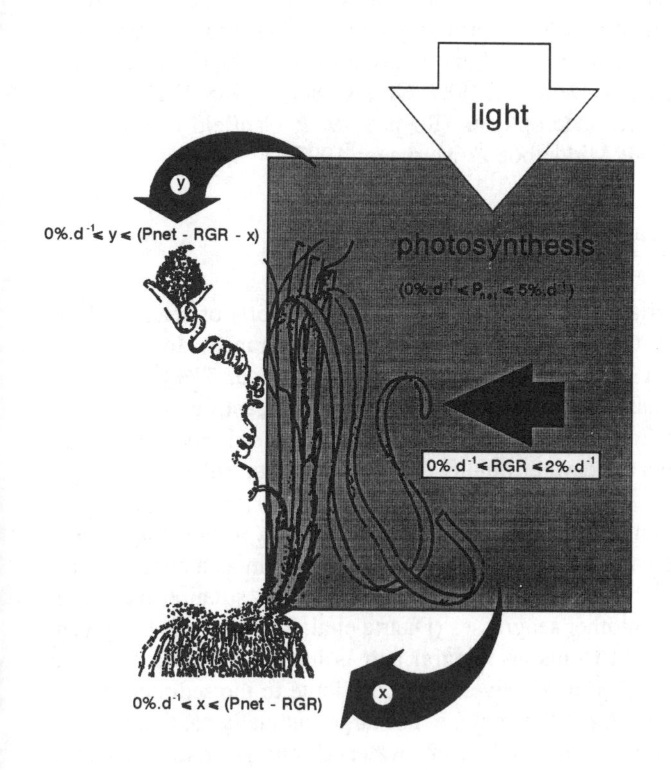

Fig. 8.2. A schematic diagram showing the general flux of carbon assimilates derived from photosynthesis ($0\%.d^{-1} \leq P_{net} \leq 5\%.d^{-1}$) in *Enhalus acoroides*. Dark arrows indicate allocations to leaf growth ($0\%.d^{-1} \leq RGR \leq 2\%\ d^{-1}$) and the production of underground biomass ($0\%.d^{-1} \leq x \leq [P_{net}$ - RGR]) and sexual reproductive structures ($0\%.d^{-1} \leq y \leq$ [Pnet-RGR - x]). RGR allocation has a priority (see Chapter 3) while allocations for x and y are yet to be fully understood.

range and (2) since RGR can not be higher than the net photosynthesis, any decrease of the latter at the lower light range should correspond to a decrease in the former.

This light-limitation mechanism (Figs. 8.1 and 8.2) appears to explain plausibly most of the other major findings in the present research. In this context, the poor correlation between net photosynthesis and relative leaf growth of *Enhalus acoroides* at the shallow and clear-water site 1a (Fig. 3.5, Chapter 3) is explained. The significant correlation between *Enhalus acoroides* flowering and available PAR (Fig. 5.5, but see also Figs. 5.6 and 5.7, Chapter 5) is consistent with the light-limitation mechanism (Figs. 8.1 and 8.2). From Fig. 5.5 (Chapter 5), it can be concluded that flowering stops at a mean PAR lower than 12 $E.m^{-2}.d^{-1}$, which closely corresponds to the PAR requirement for a mean RGR of ca. 2 $\%.d^{-1}$ (Fig 8.1). Further, this explains the rarity of *Enhalus* flowering at the deepest site (1c) which receives PAR of approximately 12 $E.m^{-2}.d^{-1}$ (Fig. 8.1; Figs. 5.5, 5.6 and 5.7, Chapter 5). It therefore appears understandable that rare *Enhalus* flowers at this site were observed when PAR level exceeded 12 $E.m^{-2}.d^{-1}$ (Fig. 5.6, Chapter 5).

Shoot densities of *Enhalus acoroides* across sites correlate negatively with differences in light availability (Fig. 4.5, Chapter 4). Apparently, *Enhalus acoroides* conforms with the general trend that plants in darker (i.e., deeper and/or more turbid) environments are larger but less dense (see for example, Bay 1984).

In summary, the above indicates that, for *Enhalus acoroides* at the present study sites, light is the most important single environmental factor determining its performance across time and space. This supports earlier discussions on the primary importance of light availability on distribution and depth limits of seagrasses (Dennison 1987; Orth & Moore 1988; Duarte 1991; Vermaat et al. in press) and freshwater macrophytes (Spence 1982; Canfield et al. 1985; Chamber & Kalff 1985; Kautsky 1988; Middelboe & Markager 1997).

PROBABLE IMPACT OF SILTATION AND/OR EUTROPHICATION

Recently, Short and Wyllie-Echeverria (1996) reviewed world-wide reports of seagrass loss and evaluated the types of causes of the seagrass declines and disappearances. In a total of 37 cases published during the period 1983-1994, 21 reported losses (or ca. 57%) were due to the reduction of "water clarity", making this category the most significant threat to global seagrass survival. Short and Wyllie-Echeverria (1996) categorized "water clarity" to include reports of nutrient loading, eutrophication, water quality, pollution, and turbidity.

The case of erosion-derived siltation in the Philippines (EMB Report 1990; Yap 1992; Vermaat et al. in press) may also be categorized as a water clarity problem as a consequence of increased turbidity and, probably, nutrient loading. In addition, heavy siltation may also modify sediment composition and/or smother seagrasses (Duarte et al. in prep). The Lingayen Gulf in which the Bolinao Reef System forms an integral part is in fact suffering from an advanced case of siltation. Among the five major rivers that contribute to the sediment loads in the Lingayen Gulf (Rivera 1997), the Agno River alone discharges annually ca. 610 million tons of suspended solids (TSS, figures derived from water discharge rate and TSS concentration values compiled by Rivera 1997).

Below, I will speculate on the implications of erosion-derived siltation for *Enhalus acoroides* populations in the area.

Nutrient loading and modification in sediment composition

This dissertation showed that shoot size of *Enhalus acoroides* is strongly affected by sediment type and nutrients. Relative to those growing at a sandy site 1a, the shoot sizes of mature *Enhalus* plants at muddy sites 5 and 6 are larger (Chapter 2). This *in situ* observation on mature plants is supported by both *in situ* and *in vitro* experiments testing the effect of sediment type on *Enhalus* seedling morphometrics (Chapter 6). Seedlings grown in sediment types 5 and 6 were larger compared with those grown in sediment types 2 and 3 (coarsest types; more than 70% of grain size > 250 μm , Table 2.1, Chapter 2).

An increased level of nutrients also strongly affected the shoot size of *Enhalus* seedlings (Chapter 6) and mature plants (Agawin et al. 1996). However, this effect on shoot size appears to be limited to nutrient increases occurring in sediment types comparable to that of site 1a (carbonate type). *Enhalus* seedlings grown in muddy type (sediment type 5), did not respond to nutrient addition (Chapter 6). This suggests that the availability of mineral nutrients in terrigenous sediments (e.g. site 5) is sufficient for *Enhalus acoroides* demands (see Erftemeijer 1993; Marschner 1986; Ericsson 1995).

While differences in sediment types and nutrient levels will result in differences in shoot size (Chapters 2 and 6), the effect of both factors on the reproductive capacity of *Enhalus acoroides* is unclear (Chapters 4 and 5). The increase in shoot size at muddy sites (5 and 6) was not paralleled with an increase in flowering (Chapter 5) and vegetative reproduction (Chapter 4). Shoot density of *Enhalus acoroides* at site 6 was higher (Fig. 4.3, Table 4.2, Chapter 4) but it is unlikely that this is due to a probable periodic increase in nutrients at the site. There exists not enough evidence supporting that nutrient increases would result in higher *Enhalus* density (see *in situ* nutrient additions, e.g., Erftemeijer 1993, in Indonesia; Agawin et al. 1996, in Bolinao). Most probably, the higher density of *Enhalus* at site 6 is a combined effect of the relatively higher survival rate of seedlings at the site (Fig. 6.2, Table 6. 1; Chapter 6) and possibly a denser seed rain which could come from the neighbouring (shallow and clear water) *Enhalus* vegetation (pers. obs.).

It can therefore be speculated that the impact of nutrient loading and modification of sediment composition (i.e., fertilization) may initially increase the shoot size of *Enhalus* growing at the nutrient-poor sites (1a, 1b, 1c, 4 and similar sites). However, in the long-term, the direct effect will cease (i.e., when the situation would already be comparable with site 5). By that time, the effect of other factors such as light limitation would be much more substantial.

Smothering by sediment

A rough estimate of the sedimentation rate in the Lingayen Gulf has been made (Santos et al. 1991 in Rivera 1997) and amounts to 4.7 $cm.yr^{-1}$. The spatial variation in this value was not given (so, certainly needs more extensive work) but values near the coast where seagrasses abound should be higher. Nevertheless, this figure (4.7 $cm.yr^{-1}$) is still within the range (2-13 $cm.yr^{-1}$, Vermaat et al. in press) which most seagrass species are expected to be able to cope with. Hence, smothering per se might just be a minor problem.

Water clarity

Speculations on the water clarity aspect of siltation can be done more quantitatively. From the approximations presented in Fig. 8.1 showing a light compensation point for net photosynthesis (LCP) of 2 $E.m^{-2}.d^{-1}$ and the threshold PAR level of 13 $E.m^{-2}.d^{-1}$ required for flowering, the impact of light climate deterioration (e.g. increase in turbidity) on the survival and depth limits of *Enhalus acoroides* can be determined. Based on an annual mean solar PAR (cloudless situation) of ca. 60 $E.m^{-2}.d^{-1}$ (Fig. 2.11, Table 2.4, Chapter 2) and an annual mean cloud attenuance of 48.1% (Table 2.4, Fig. 2.11, Chapter 2), the critical survival and flowering depths can be calculated given hypothetical turbidity values (e.g., K_d range: 0.3-6.0 m^{-1}, Fig. 8.3). It is predicted (Fig. 8.3) that *Enhalus acoroides* survives at very clear waters (K_d ca. 0.3 m^{-1}) as deep as 9 m. This prediction suggests that at site 3 (K_d ca. 0.3 m^{-1}; depth ~ 5 m) where *Enhalus acoroides* is not present, this species should be able to survive. The findings that both seedlings (Chapter 6) and mature transplants (Chapter 7) survived at the site are consistent with the above prediction. Coles et al. (1987) also reported *Enhalus acoroides* to occur at 5 m in the waters of NE Queensland (Australia). For flowering, the depth limit at clear waters is ca. 3 m (Fig. 8.3).

If water turbidity (K_d) increases, for instance, from 0.3 to 1 m^{-1}, the critical depths for survival and flowering decreases to 2.7 m and 0.9 m respectively. This would mean that *Enhalus*

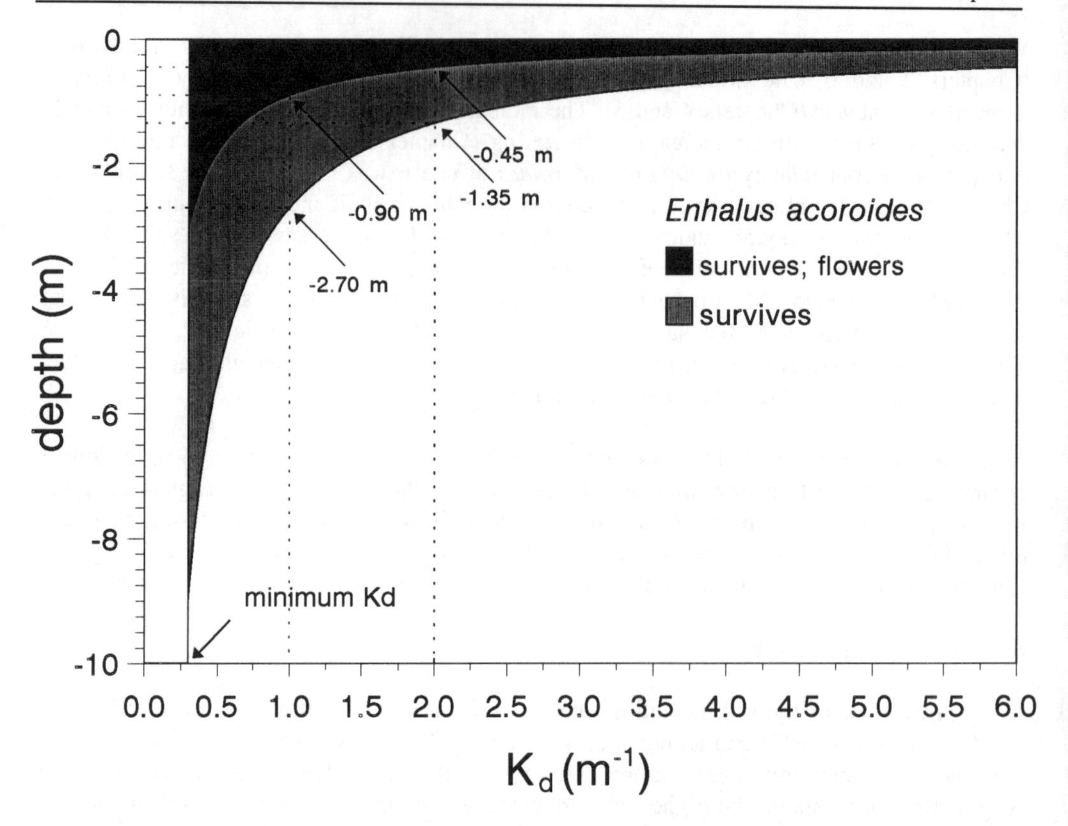

Fig. 8.3. Water depth (m, smaller ticks = 0.25 m) and turbidity (K_d, m^{-1}) conditions at which *Enhalus acoroides* survives (black and grey shades) and flowers (black shade). To illustrate the impact of a higher K_d (for example, $K_d = 1\ m^{-1}$ and $K_d = 2\ m^{-1}$), the corresponding critical depths (horizontal dotted lines) for *Enhalus acoroides* to survive and/or produce flowers are indicated. The mean K_d at a very clear water site (1a) is also shown. It is the observed minimum K_d over the whole study period.

acoroides at, for example, site 4 (depth ca. 1 m; K_d ca. 0.4 m^{-1}; where at present conditions, flowering intensity is high; Chapter 5), will most probably no longer be able to flower. Further increase in K_d to 2 m^{-1} predicts that *Enhalus acoroides* will not survive at sites deeper than 1.35 m while *Enhalus* inhabiting deeper than 0.45 m will not be able to produce flowers.

Per se, Fig. 8.3 can be useful as quick reference to speculate upon the effect of the increase in water turbidity on *Enhalus* shoot density and floral distributions. However, with the existing high sediment discharges of the major rivers in Lingayen Gulf, Rivera (1997) predicted that only minimal effect would be experienced by seagrasses in Bolinao as most TSS is transported towards the east of the gulf. Still, *Enhalus* populations in the southern and eastern coasts of Lingayen Gulf (especially where predicted TSS > 25 mg/l, i.e., $K_d > 1\ m^{-1}$; Rivera 1997, Fig. 7.32) are then probably stressed. For the Bolinao area, Rivera (1997) predicted that even if the TSS concentration in the discharge water of the biggest river (Balingasay River) in the area is increased ca. 6-fold (i.e., from 17 mg/l to 100 mg/l), the effect on the TSS concentration in the nearby coastal waters would still be minimal. However, a TSS concentration of 100 mg/l (as used in model calculations by Rivera 1997) in Balingasay River means that the erosion rate of the river catchment area (72 km^2) was increased from the existing 5 $tons.ha^{-1}.yr^{-1}$ to ca. 30 $tons.ha^{-1}.yr^{-1}$. This value is, by far, lower than that of the Agno River catchment area (1000 $tons.ha^{-1}.yr^{-1}$). This thus basically means that if the remaining "forests" in the Balingasay River basin are destroyed (e.g., due to construction of a cement plant) to a situation comparable with the Agno catchment, then the eventual turbidity of the waters nearby (esp. the southwestern part during SW moonsoon; Fig. 9.7, Rivera 1997) could be far higher.

To summarize, this section showed that: (1) erosion-derived siltation is an existing problem within a broader regional (Lingayen Gulf) perspective; (2) although increased nutrient loading, modification of sediment composition and smothering by sediment may all have direct effects on *Enhalus acoroides*, the effect of the reduction in water clarity is probably more substantial; (3) *Enhalus* populations at the southeastern coasts of Lingayen Gulf would probably have to cope with poor light availability; and (4) although at the moment, siltation is, in general, not likely stressing *Enhalus* populations in Bolinao, further destruction of the remaining upland "forests" could pose serious threats (i.e., by reducing light) to seagrass populations in the area especially those inhabiting deeper waters.

FUTURE PERSPECTIVES

This dissertation has succeeded in gaining more insights into the biology of *Enhalus acoroides* but more questions remain:

1. *Genetic diversity.* For seagrasses, like any organism, to survive in a changing environment (e.g. light climate deterioration), they must be able to acclimate (i.e., phenotypic plasticity) and/or adapt through genetic recombination. For the latter, there is hardly any information available. There is therefore a need to do research along this line.

2. *Fate of seeds.* From Chapter 5, it was estimated that the areal seed output could be substantial (93 $seeds.m^{-2}.yr^{-1}$) but the number of seedlings which settled in the permanent quadrats was comparatively less (0.67 seedlings. m^{-2} $.yr^{-1}$). An information gap exists as to the fate of *Enhalus* seeds.

3. *Seagrass dynamics at the threshold PAR levels.* Site 1c could be the best study area for this theme. For *Enhalus*, this site is located near the threshold PAR level required for flowering. Mature transplants within the same site (1c, Chapter 7) did not establish, implying a delicate balance in its shoot-root system. For the dynamics of other seagrass species, e.g., *Thalassia hemprichii* and *Cymodocea serrulata* which depth limits seem to oscillate (pers. obs., i.e., relative to the permanent markers at the site), this site is an interesting study area.

4. *Inhibitory factor at site 2.* The unknown factor (or combinations thereof) causing the immediate (within weeks) die-off of both *Enhalus acoroides* seedlings and mature transplants at site 2 (Chapter 6; Chapter 7), is undoubtedly a relevant consideration in explaining the local and regional distribution of *Enhalus acoroides*. Results from this study discriminated, at least, light, sediment type and nutrient factors. This line should be a worthwhile component of future studies.

5. *Plasticity of PI-curve parameters*. The accuracy of the calculated photosynthesis (e.g. daily summations, Chapter 2) can be further refined by obtaining independent spatial and temporal estimates of PI-curve parameters. Across populations of *Enhalus acoroides* at different depths with similar water clarity and sediment characteristics , PI-curve parameters (i.e., P_{max} and K_m, R; see Chapter 3) may not vary (Rollon & Fortes, unpubl. data). Temporal differences in these parameters were also unclear (Flores & Fortes 1992; Vermaat et al. 1995), but are significant for temperate seagrasses (Vermaat & Verhagen 1996).

6. *Potential interactions between nutrients, photosynthesis and flowering*. In his review, Ericsson (1995) concluded that the deficiency of major growth-limiting nutrients N, P and S affects the synthesis of new tissues but does not affect carbon fixation (i.e., rate of photosynthesis) itself. Hence, N-P-S additions may not result in significant differences in P_{max} and K_m. However, there is evidence that nutrient (N-P-K) addition results in slightly higher P_{max} and K_m values (Agawin et al. 1996). This effect could be due to potassium (in the form of K_2O; Agawin et al. 1996; this study, Chapter 6) which has been shown to directly affect the rate of photosynthesis (Marschner 1986; Ericsson 1995). This implies that, as here presented (Chapter 3), estimates of net photosynthesis of *Enhalus* at site 6 (and probably also site 5; see K contents in sediments, Table 2.3, Chapter 2) may be slightly underestimated. This does not change the temporal pattern as presented but could further elucidate the energy budget of plants inhabiting at these sites. Looking at RGR-P_{net} ratios (Fig. 3.5, Chapter 3), *Enhalus* at sites 5 and 6 suffer similar resource limitation as those found at site 1c. On this basis, it is rather difficult to explain the significantly higher *Enhalus* flowering at sites 5 and 6 than at site 1c (Figs. 5.3, 5.4; Chapter 5). Indeed, the calculated net photosynthesis at sites 5 and 6 might have been slightly underestimated. This certainly merits further investigation.

6. *Periphyton colonizing artificial substrates vs. natural leaves*. Light attenuance of periphyton colonizing on artificial substrate incubated at different sites has been measured in this research and a general implication has been speculated (Table 2.4, Chapter 2). However, the incorporation of this aspect on the photosynthesis calculation (Chapter 3; and also Fig. 8.3) has been postponed due to apparent differences in quantity (and probably quality; data in progress) of periphyton colonizing artificial substrates and natural leaves of *Enhalus acoroides* (pers. obs.), the latter having visibly less. From Table 2.4 (Chapter 2), it was shown that at site 3, the mean PAR available at seagrass depth is ca. 1.7 $E.m^{-2}.d^{-1}$ which, according to Fig. 8.1, means that it is not possible for *Enhalus* to survive in this condition. The fact that seedlings and mature transplants survived at the site (Chapter 6; Chapter 7) suggests that PAR availability should be higher than 2.13 $E.m^{-2}.d^{-1}$ which can be realized if PAR attenuance due to periphyton can be less (for example, PAR would be ca. 6.64 $E.m^{-2}.d^{-1}$ disregarding periphyton effect).

Therefore, there is a need to first establish the relation between the values obtained from the artificial substrates and those obtained from natural leaves. An equally worthwhile effort is to develop techniques of directly measuring light attenuance of periphyton on *Enhalus* leaves (e.g. Bulthuis & Woelkerling 1983 for *Heterozostera tasmanica*) rather than on artificial substrates. But initial efforts by this research towards this line proved difficult as the leaves of *Enhalus* are too thick to allow light to pass through.

The list above indicates that there is still so much work to do. However, with the amount of attention given to Philippine seagrasses growing tremendously over the last decade (mainly through international cooperation, e.g. ASEAN-Australia, EU and WOTRO-funded projects), I am optimistic that, very soon, studies addressing these lines can be carried out. To refine our understanding of Philippine multi-species seagrass meadows, it is obvious that similar studies investigating the life strategies of other seagrass species are necessary. Next in priority would be *Thalassia hemprichii* which, in some seagrass meadows, is even more abundant than *Enhalus acoroides*. In this research, some aspects of *Thalassia* growth (Chapter 3), density and recolonization (Chapter 4) have been addressed. However, aspects of sexual reproduction and environmental effect studies are scarce. Even less available are studies addressing similar lines for other seagrasses commonly found in Bolinao, Pangasinan. These other species include *Syringodium isoetifolium*, *Halophila ovalis*, *Halodule uninervis*, *Cymodocea rotundata* and *Cymodocea serrulata*.

LITERATURE CITED

Agawin, N.S.R., C.M. Duarte and M.D. Fortes. 1996. Nutrient limitation of Philippines seagrasses (Cape Bolinao, NW Philippines): *in situ* experimental evidence. Mar. Ecol. Prog. Ser. 138: 233-243.

Bulthuis, D.A. and W.J. Woelkerling. 1983. Biomass accumulation and shading effects of epiphytes on leaves of the seagrass, *Heterozostera tasmanica*, in Victoria, Australia. Aquat. Bot. 16: 137-148.

Canfield, Jr. D.E., K.A. Langeland, S.B. Linda and W.T. Haller. 1985. Relations between water transparency and maximum depth of macrophyte colonization in lakes. J. Aquat. Plant Manage. 23: 25-28.

Chambers, P.A. and J. Kalff. 1985. Depth and biomass distribution of submerged aquatic macrophyte communities in relation to Secchi depth. Can. J. Fish. Aquat. Sci. 42: 701-709.

Coles, R.G., W.G. Lee Long, B.A. Squire, L.C. Squire, and J.M. Bibby. 1987. Distribution of seagrasses and associated juvenile commercial penaeid prawn in northeastern Queensland waters. Aust. J. Mar. Freshwater Res. 38: 103-119.

Dennison, W.C. 1987. Effects of light on seagrass photosynthesis, growth and depth distribution. Aquat. Bot. 27: 15-26.

Duarte, C.M. 1991. Seagrass depth limits. Aquat. Bot. 40: 363-377.

Duarte, C.M., J. Terrados, N.S.R. Agawin, M.D. Fortes, S. Bach and W.J. Kenworthy. In prep. Response of a mixed Philippine seagrass meadow to experimental burial. To be submitted to Mar. Ecol. Prog. Ser.

EMB Report. 1990. The Philippine environment in the eighties. Environmental Management Bureau, Quezon City, Philippines, 302 pp.

Erftemeijer, P.L.A. 1993. Factors limiting growth and production of tropical seagrasses: nutrient dynamics in Indonesian seagrass beds. PhD Dissertation, Katholieke Universiteit Nijmegen, The Netherlands, 173 pp.

Ericsson, T. 1995. Growth and shoot:root ratio of seedlings in relation to nutrient availability. Plant and Soil 168-169: 205-214.

Flores, G.A. and M.D. Fortes, M.D. 1992. Photosynthetic-respiratory responses of seagrasses in Pislatan, Cape Bolinao, Pangasinan. UNDP-PCAMRD: Proc. 2nd Symp. Phil. Assoc. Mar. Sci., Tawi-Tawi, Philippines.

Ingestad, T. and G.I. Ågren. 1995. Plant nutrition and growth: basic principles. Plant and Soil 168-169: 15-20.

Gifford, R.M. and L.T. Evans. 1981. Photosynthesis, carbon partitioning, and yield. Ann. Rev. Plant Physiol. 32: 485-509.

Kautsky, L. 1988. Life strategies of aquatic soft bottom macrophytes. Oikos 53: 126-135.

Marschner, H. 1986. Mineral nutrition of higher plants. Academic Press, London, 674 pp.

Middelboe, A.L. and S. Markager. 1997. Depth limits and minimum light requirements of freshwater macrophytes. Freshwat. Biol. 37: 553-568.

Orth, R.J. and K.A. Moore. 1988. Distribution of *Zostera marina* L. and *Ruppia maritima* L. *sensu lato* along depth gradients in the lower Chesapeake Bay, U.S.A. Aquat. Bot. 32: 291-305.

Rivera, P.C. 1997. Hydrodynamics, sediment transport and light extinction off Cape Bolinao, Philippines. PhD Dissertation, IHE-WAU, The Netherlands, 244 pp.

Short, F.T. and S. Wyllie-Echeverria. 1996. Natural and human-induced disturbance of seagrasses. Environmental Conservation 23: 17-27.

Short, F.T., D.M. Burdick and J.E. Kaldy III. 1995. Mesocosm experiments quantify the effects of eutrophication in eelgrass, *Zostera marina*. Limnol. Oceanogr. 40: 740-749.

Spence, D.H. 1982. The zonation of plants in freshwater lakes. Adv. Ecol. Res. 12: 37-126.

Vermaat, J.E. and M.J.M Hootsmans. 1994. Growth of *Potamogeton pectinatus* L. in a temperature-light gradient. In: W. van Vierssen, M.J.M. Hootsmans and J.E. Vermaat (eds.). Lake Veluwe, a macrophyte-dominated system under eutrophication stress. Geobotany 21. Kluwer, Netherlands, 40-61 pp.

Vermaat, J.E. and F.C.A. Verhagen. 1996. Seasonal variation in the intertidal seagrass *Zostera noltii* Hornem.: coupling demographic and physiological patterns. Aquat. Bot. 52: 259-281.

Vermaat, J.E., C.M. Duarte and M.D. Fortes. 1995. Latitudinal variation in life history patterns and survival mechanisms in selected seagrass species, as a basis for EIA in coastal marine ecosystems (Final Report). Project EC DG XII-G CI1*-CT91-0952, IHE Delft, The Netherlands, 38 pp. + 2 annexes.

Vermaat, J.E., N.S.R. Agawin, C.M. Duarte, S. Enriquez, M.D. Fortes, N. Marba, J.S. Uri and W. Van Vierssen. In press. The capacity of seagrasses to survive eutrophication and siltation, across-regional comparisons. Ambio.

Went, F.W. and L.O. Sheps. 1969. Environmental factors in regulation of growth and development: ecological factors. In: F.C. Steward (ed). Plant physiology- a treatise, Vol. 5a. Academic Press, NY, 229-406 pp.

Summary

Seagrasses can form extensive and highly productive meadows that provide food, shelter and nursery grounds for various organisms, many of which are commercially important. In many human settlements along the coasts, seagrass meadows comprise a considerable component of the resource base for livelihood. However in recent years, reports of physical losses of seagrasses are growing. Topping the list of causes of such losses is the deterioration of water clarity associated with human-induced disturbances, such as siltation and eutrophication. The mechanism behind these empirical observations is complex and our understanding of such mechanism is, until today, limited.

In the tropical Indo-West Pacific region, *Enhalus acoroides* (L.f.) Royle is the largest among the total of ca. 20 seagrass species occurring in the coastal waters. This species is the most widely distributed and structurally dominates many seagrass meadows. Thus, understanding the biology of *Enhalus* could contribute considerably to the understanding of how seagrass meadows would respond to environmental changes such as light deterioration. In Bolinao (NW Philippines), the coral reef system is largely comprised of a reef flat where seagrasses abound, often in multi-species meadows but also in monospecific vegetations of *Enhalus acoroides*. Choosing Bolinao as the study area, this dissertation approached the general question above by following three main research lines: (1) describing growth characteristics of *Enhalus* in the established phase; (2) explaining such characteristics in relation to prevailing environmental conditions; and (3) determining the response of propagules to manipulated environmental changes. The execution of these research lines has been carried out through a combination of *in situ* and semi-controlled laboratory studies.

To summarize, this dissertation demonstrates that variability in light per se explains most of the variation in several aspects in the growth characteristics of *Enhalus acoroides*. Across a spatial light gradient, this species was found to acclimate to darker environments by increasing its shoot size. Over time, the variation in relative leaf growth (RGR) correlated strongly with the variation in light availability. In this respect, calculations relating net photosynthesis with light further revealed that the variation in light due to cloud cover accounted for more than 85% of the variation in net photosynthesis of *Enhalus acoroides* in all study sites. But while RGR strongly varies temporally, it remains relatively constant across sites. This characteristic means that *Enhalus* in darker environments (i.e., where net photosynthesis is lower) has to limit its energy allocation to other sink activities such as below-ground production and/or flowering. This hypothesis has been confirmed partially by the study on *Enhalus* flowering. In that chapter, it was shown that both the intensity and frequency of flowering strongly correlate with light availability. The estimated threshold light level (ca. 12 $E.m^{-2}.d^{-1}$) required for flowering plausibly explains the rarity of flowers observed in deeper *Enhalus* populations not receiving this required quantity of light. For successful fruiting, a necessary condition is successful pollination. This was only found possible in shallow sites.

The manipulative experiments conducted in this research affirmed the primary importance of light in the biology of *Enhalus acoroides. In situ* and semi-controlled light manipulations on *Enhalus* seedling cultures yielded larger seedlings in darker environments. This finding confirms the earlier hypothesis that the *in situ* shoot size distribution of *Enhalus acoroides* (i.e., being larger in darker environments) is an acclimation to light during the development stage. The results of shading and transplantation experiments on the mature stands of *Enhalus acoroides* are also consistent with earlier results relating shoot size and growth of mature *Enhalus* with temporal variation in light. Short-term imposed 70% shading *in situ* significantly reduced shoot size and leaf growth of *Enhalus acoroides.* In manipulating light availability by transplantation, this research shows that *Enhalus acoroides* performs better when transplanted from dark to clear environments than vice versa. Combining these manipulative experiments (Chapters 6 and 7) with *in situ* spatial differences in shoot size (Chapter 3), it can be postulated that the size variation observed in *Enhalus* populations in Bolinao is largely a phenotypic response to a varying environment.

By physical creation of gaps, this dissertation predicts that in a shallow, clear-water multi-species meadow, *Enhalus acoroides* could recolonize (by vegetative means) completely a clearance size of 0.25 m^{-2} only after 10 years. In this experiment, the importance of recruitment by sexual propagules (seedlings) into the gaps was found to be very low, although the areal seed output of the site was estimated to be high (93 seeds.m^{-2}.yr^{-1}). Relative to *Enhalus*, the co-dominant species, *Thalassia hemprichii*, was found to fully recover much faster, i.e., after ca. 2 years.

In view of the foregoing results, the impact of erosion-derived siltation which is threatening the study area is predicted. A biplot relating turbidity (K_d) with depth maxima of *Enhalus* survival and flowering is presented. The plot predicts that, in clear waters (Kd = 0.3 m^{-1}), *Enhalus* could colonize sites up to 9 m deep. If K_d increases to 1 m^{-1}, no *Enhalus* population could be expected to occur deeper than 3 m. Parallel predictions are made for flowering, which in clear waters could occur up to 3 m but could not be expected to occur deeper than 1 m if K_d increases to 1 m^{-1}. In this view, it can be predicted that an increase in water turbidity (e.g. siltation and/or eutrophication) would directly result in a decrease in the survival and flowering depth limits of *Enhalus acoroides.*

Sa laktud

Ang mga lusay makahimo pagmugna ug lapad ug produktibong lasang sa dagat nga nagpakaon, nagtaming, ug nagmatutu sa nagkadaiyang matang sa organismo, kadaghanan niini adunay komersiyanhong bili. Sa daghang tawhanong katilingban nga nagpuyo sa kabaybayunan, dakong bahin ang kalusayan sa tinubdan sa panginabuhi. Apan niniing pipila ka miaging mga tuig, nagkabaga ang mga taho mahitungod sa pagkahanaw sa kalusayan sa daghang dapit sa tibuok kalibutan. Nag-una sa mga hinungdan mao ang pagkalubog sa tubig gumikan sa tawhanong kakulian sama sa pagkankan sa naupaw nga kabukiran ug yutropikasyon. Ang mekanismo luyo niining maong kalambigitan lawom sabton ug ang atong kahibalo sa maong mekanismo, hangtud karon, taphaw.

Sa tropikanhong rehiyon taliwala sa kadagatang Indiano ug Kasadpang Pasipiko, ang *Enhalus acoroides* (L.f.) Royle mao'y kinadak-an sa mokapin kun kulang 20 ka matang sa lusay nga makita niining bahina sa kalibutan. Kining matanga maoy sagad makita ug mihawod sa daghang kalusayang dapit. Tungod niini, ang pagsabot sa kinaiyahan sa *Enhalus*, sa walay pagduda, makatampo ug dako alang sa pagtubag sa pangutana kun unsaon pagsugakud sa kalusayan ang pagbag-o sa palibot sama sa pagkunhod sa kahayag (pagngi-ob). Sa Bolinao (Amihanang Kasadpan sa Pilipinas) nga gipiling dapit niining maong pagtuon, dakong bahin ang hunasan sa makaplagang *Coral Reef system* (kagusuan). Niining maong hunasan, dasuk ang kalusayan nga sagad nagsagol ang daghang matang, o dili ba, mga tanaman sa nag-inusarang *Enhalus acoroides*. Niining maong pagtuon, gitumong tubagon ang nahaunang pangutana (sa ibabaw) pinaagi sa pagdukiduki sa mosunod nga tulo ka nga mga sumbanan: (1) ang paghulagway sa mga kinaiyahan sa establisado nang *Enhalus*; (2) ang paglambigit niining maong mga kinaiyahan tali sa mga kahimtang sa iyang palibot; ug (3) ang pagsayod sa reaksyon sa mga binhi sa tinuyong pagbag-o sa kahimtang sa palibot. Kining maong mga pagsusi gimatuod pinaagi sa paghimo ug mga eksperimento didto sa mismong kalusayan ug usab mga pagtuon sulod sa laboratoryo.

Sa paglaktud, gipamatud-an niining maong pagtuon nga ang kausaban sa kahimtang sa kahayag maoy lig-ong hinungdan sa kausaban sa daghang aspeto sa kinaiyahan sa *Enhalus acoroides*. Sa pagtandi sa mga kausaban latas sa lainlaing populasyon, kining matanga sa lusay nagdangop sa nagkangi-ob nga palibot pinaagi sa pagbaton ug dagkong usbong. Latas sa lainlaing panahon, ang relatibong pagtubo sa dahon (RPD) adunay lig-ong kalambigitan sa pag-usab-usab sa gikusgon sa natagamtamang kahayag. Sa niining kalabutan, ang paglambigit sa *photosynthesis* ug kahayag nagbutyag nga ang pag-usab-usab sa gikusgon sa kahayag gumikan sa kapanganurun maoy hinungdan sa mokapin sa 85% sa kausaban sa *photosynthesis* sa *Enhalus acoroides* sa tanang gisusing dapit. Apan samtang ang RPD nag-usab-usab duyog sa kausaban sa panahon, kini nagpabiling pareho tandi sa nagkalainlaing populasyon. Kining maong kinaiyahan nagpasabot nga ang *Enhalus* sa mangi-ob nga puluy-anan (diin diyutay ang abot sa *photosynthesis*), nagkinahanglan nga ikunhud ang pahat ngadto sa ubang bahin sa pagpakabuhi sama sa pagpamulak, pagpamunga ug pagpanggamot. Kining maong pangagpas giuyunan sa detalyadong patuon sa pagpamulak sa *Enhalus*. Sa maong yugto (Kapitulo 5), gipamatud-an nga adunay lig-ong kalambigitan taliwala sa pagpamulak ug sa ang-ang sa natagamtamang kahayag. Ang nabanabana nga kina-ubsang gikusgon sa kahayag (12 $E.m^{-2}.d^{-1}$) nga gikinahanglan alang sa pagpamulak, hugot nga nagpatin-aw ngano nga sa kinalaluman ug kinangi-oban nga populasyon sa *Enhalus* taghap kaayong panghitabo ang pagpamulak. Alang sa malampusong pagpamunga, gikinahanglan ang malampusong *pollination*. Kini mahimo lamang sa mabaw nga lugar.

Ang mga gihimong eksperimento nga mituyo sa pag-usab sa kahimtang sa palibot mipamatuod usab sa kamahinungdanon sa kahayag alang sa kinabuhi sa *Enhalus acoroides*. Ang mga eksperimentong nagmanipula sa kahayag didto mismo sa kalusayan ug usab sa mga pagtuon sa laboratoryo mipadayag nga dagko ang mga binhing gipatubo sa sinilungang kahimtang. Kun buot sabton, kining maong nakaplagang kasayuran mipalig-on sa nahaunang pangagpas nga ang kalainan sa gidak-on sa usbong sa *Enhalus* latas sa lainlaing dapit (nga sa ato pa, dagko sa ngi-ob nga palibot) usa ka kinaiyahan sa maong lusay sa pagdangop sa ngi-ob nga kahimtang samtang kini nagadako. Sa laing bahin, ang mga reaksyon sa mga hingkod nga *Enhalus* sa tinuyong pag-usab sa kahayag nahiuyon usab sa nahaunang mga resulta nga naglambigit sa gidak-on sa usbong ug RPD. Ang tinuyong 70 porsiyentong pag-us-us sa kahayag pinaagi sa pagpandong misangpot sa dakong pagkunhod sa gidak-on ug tinubuan sa *Enhalus acoroides*. Ang tinuyong pag-usab sa kahimtang sa kahayag pinaagi sa pagluka sa mga hingkod ug pagbalhin niini sa laing dapit, tataw nga mipadayag nga kadtong mga binalhin gikan sa ngi-ob nga dapit paingon sa tin-aw nga lugar, miarang-arang ang kahimtang sa mga tanom. Kun lantawon kining maong mga resulta (mga Yugto 6 ug 7) uban sa kalainan sa usbong nga makita sa kinaiyahan (Yugto 3), mahimong maingon nga ang kausaban sa gidak-on sa *Enhalus* sa lainlaing populasyon sa Bolinao, sa dakong bahin, usa ka reaksyong *phenotypic* sa nag-usab-usab nga kahimtang sa palibot.

Pinaagi sa tinuyong pagmugna ug upaw nga luna, kining maong pagtuon mitagna nga, sa mabaw, tin-aw ug sagol nga kalusayan, makahimo ang *Enhalus acoroides* pagbalik sa hingpit (pinaagi sa pagpanalingsing; gidak-on sa hawan = 0.25 m^2) sulod sa 10 ka tuig. Niining maong pagsusi, gipadayag nga dili dako ang papel sa sekswal nga mga binhi sa pagbalik pagtungha sa *Enhalus*, bisan pa man nga ubay-ubay ang gibanabana nga kadaghanon sa produksyon sa mga liso niining dapita (93 ka liso kada metro kwadrado matag tuig). Kun itandi sa *Enhalus*, ang usa usab ka dominanteng matang sa lusay, *Thalassia hemprichii*, nakahimo pagbalik sa hingpit sulod lamang sa duha ka tuig.

Sigun sa nahaunang saysay sa mga resulta, kining maong pagtuon milantaw sa mahimong sangputanan sa nanghulgang suliran (pagkalubog gumikan sa bunbun nga dala sa baha gikan sa padayon nga pagkaupaw sa kabukiran). Gipakita ang yanong kalambigitan sa *turdidity* (K_d, sukod sa kalubugon sa tubig) ug ang kalalumon sa tubig nga mabuhi pa ug mamulak ang *Enhalus*. Gitagna nga sa tin-aw nga tubig (K_d = 0.3 m^{-1}) mahimong mabuhi pa ang *Enhalus* hangtod sa kalalumon nga 9 baras. Kun ang K_d mosaka sa 1 m^{-1}, nan, wala nay *Enhalus* nga malauman sa kalalumon nga 3 baras. Gihimo usab ang samang pagpanagna alang sa pagpamulak nga, sa tin-aw'ng tubig, mahimong manaa sa kalalumon nga 3 baras apan dili na malauman nga maanaa pa sa kalalumon nga 1 bara kun ang K_d mosaka ngadto sa 1 m^{-1}. Sa niining paglantaw, mabanabana nga ang pagkalubog sa tubig (pinaagi pananglit sa *siltation* o *eutrophication*) tul-id nga mosangpot sa pagkunhod sa limit sa kalalumon nga diin mabuhi ug mamulak ang *Enhalus acoroides*.

Curriculum vitae

René Nadál Rollón was born on 12 August 1966 in Bohol, The Philippines. He finished secondary school in 1983 at the Zamboanga del Sur National High School, Pagadian City. In the same year, he started his college education at the University of the Philippines where he got a Bachelor of Science in Fisheries degree in 1987. For about half a year, he worked as a University Research Assistant at the Institute of Fisheries Development and Research (IFDR) of the College of Fisheries. The research project at IFDR was on formulating food products out of extracts from seaweeds *Gracilaria* and *Hydroclathrus*. His involvement in ecological research started in 1988 when he joined the Marine Science Institute of the College of Science, University of the Philippines (as a University Research Assistant and, later, as a University Research Associate). Mostly, he was involved (1988-1992) in the seagrass research component of the ASEAN-Australia Project on Coastal Living Resources under the guidance of Dr. Miguel D. Fortes. He was also partly involved in the IDRC subproject (also led by Dr. M.D. Fortes) on the Inventory and Stock Assessment of economically important seaweeds in the Philippines. While about to finish his Masters degree in Marine Biology (1989-1992) at the College of Science, he was granted a Dutch fellowship in 1992 to pursue both MSc and PhD degrees in an integrated and sandwich scheme. As part of his Masters degree in Environmental Science and Technology (IHE-WAU, 1992), he joined a microcosm experiment and wrote an MSc thesis on the effects of nutrient and insecticide additions on the macrophyte *Elodea nuttallii*. This PhD dissertation integrates several studies that investigate the response of the seagrass *Enhalus acoroides* (L.f.) Royle to environmental changes.

PUBLICATIONS

Brock, T.C.M., R.M.M. Roijackers, **R.N. Rollon**, F. Bransen and L. Vanderheyden. 1995. Effects of nutrient loading and insecticide application on the ecology of *Elodea*-dominated freshwater microcosm II. Responses of macrophytes, periphyton and macroinvertebrate grazers. Arch. Hydrobiol. 134(1):53-74.

Pamintuan, I.S., P.M. Aliño, E.D. Gomez and **R.N. Rollon**. 1994. Early successional patterns of invertebrates in artificial reefs established at a clear and silty areas in Bolinao, Pangasinan, Northern Philippines. Bull. Mar. Sci. 55(2-3): 867-877.

Rollon, R.N. 1992. Effects of nutrients (N,P) and insecticide Dursban® 4E on the macrophyte *Elodea nuttallii* in a ditch microcosm experiment. Masters Thesis, IHE-WAU, The Netherlands

Rollon, R.N. and M.D. Fortes. 1990. Growth rates and production of *Enhalus acoroides* (L.f.) Royle and seagrass zonation in Bolinao, Pangasinan. pp. 17-30, In: A.C. Alcala and L.T. McManus (eds.). Proc. First National Symp. Marine Science, Bolinao, Pangasinan, Philippines.

Rollon, R.N. and M.D. Fortes. 1991. Structural affinities of seagrass communities in the Philippines. pp. 333-346, In: A.C. Alcala (ed.). Proc. Regional Symp. Living Coastal Resources. Manila City, Philippines.

Rollon, R.N. and M.D. Fortes. 1992. Applicability of the plastochrone interval method in estimating the aboveground production of *Enhalus acoroides* (L.f.) Royle. In: L.M. Chou and C.R. Wilkinson (eds.). Proc. of the Third ASEAN Science and Technology Week Conference (Marine Science: Living Coastal Resources), Singapore.

Rollon, R.N., J.P. Tiquio and M.D. Fortes. 1992. Variability in the estimated minimum number of quadrats yielding the asymptote value of diversity in different seaweed communities, pp. 209-222, In: H.P. Calumpong and E.G. Meñez (eds.). Proc. of the 2nd RP-USA Phycology Symposium/Workshop, January 1992, Cebu, Philippines.

Rollon, R.N. and M.D. Fortes. 1993. Spatial and temporal variability in growth rate and production of *Enhalus acoroides* (L.f.) Royle at Bolinao reef flat, Pangasinan, Philippines. University of Guam Press, UOG Station Mangilao, GU 96923.